the AUGUST GALES

THE TRAGIC LOSS of FISHING SCHOONERS in the NORTH ATLANTIC, 1926 and 1927

GERALD HALLOWELL

Gerald Hallowell
To David
Best wishes

NIMBUS PUBLISHING

Copyright © 2013, Gerald Hallowell

All rights reserved. No part of this book may be reproduced, stored in a retrieval system or transmitted in any form or by any means without the prior written permission from the publisher, or, in the case of photocopying or other reprographic copying, permission from Access Copyright, 1 Yonge Street, Suite 1900, Toronto, Ontario M5E 1E5.

Nimbus Publishing Limited
3731 Mackintosh St, Halifax, NS B3K 5A5
(902) 455-4286 nimbus.ca

NB1055

Printed and bound in Canada

Author photo: Tina Loo
Interior design: Troy Cole, Envision Graphic Design
Cover design: Heather Bryan
Cover artwork: *The August Gales* by Jay Langford

Excerpts from *Down to the Sea: The Fishing Schooners of Gloucester* by Joseph Garland Reprinted by permission of David R. Godine, Publisher, Inc. Copyright © 1983 by Joseph Garland

Library and Archives Canada Cataloguing in Publication

Hallowell, Gerald, author
The August gales : the tragic loss of fishing schooners in the North Atlantic, 1926 and 1927 / Gerald Hallowell.

Includes bibliographical references.
Issued in print and electronic formats.
ISBN 978-1-77108-046-0 (pbk.).—ISBN 978-1-77108-047-7 (pdf).
—ISBN 978-1-77108-049-1 (mobi).—ISBN 978-1-77108-048-4 (epub)

1. Fisheries—Accidents—Atlantic Coast (Canada). 2. Shipwrecks—Atlantic Coast (Canada). 3. Fishers—Nova Scotia. 4. Fishers—Newfoundland and Labrador. 5. Storms—Atlantic Coast (Canada). I. Title.

VK1275.A856H35 2013   363.11'96392209715   C2013-903446-3
                                                  C2013-903447-1

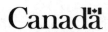 

Nimbus Publishing acknowledges the financial support for its publishing activities from the Government of Canada through the Canada Book Fund (CBF) and the Canada Council for the Arts, and from the Province of Nova Scotia through the Department of Communities, Culture and Heritage.

CONTENTS

Dark Days of Mourning 1
1 Lunenburg and the Banks Fishery 7
2 An Island in a Stormy Sea 39
3 The August Gale of 1926 61
4 The August Gale of 1927 91
5 Newfoundland and the August Gale of 1927 125
6 The Gloucester Connection 161
7 The Call of the Sea 197
Afterword: Days of Festivity and Hope 229

Acknowledgements 233
Notes 235
Bibliography 241
Image Credits 247
Index 248

The *Uda R. Corkum* racing in 1921.

DARK DAYS *of* MOURNING

UNDER A BRILLIANT BLUE LUNENBURG SKY, the vessels of the fleet riding calmly at anchor in the placid waters of the harbour, it was difficult to imagine the horrors the men had faced in the raging, roiling seas off Sable Island. The brutal North Atlantic storm, on August 7–8, 1926, had sent fifty men to a watery grave.

Some six or seven thousand people had gathered in the small town on Nova Scotia's South Shore in early October to mourn the loss of two banks schooners, the *Sylvia Mosher* and the *Sadie A. Knickle*, vanished forever beneath the waves with all hands on board. Men, women, and children arrived in six hundred automobiles, and more people came by sailboat, rowboat, and motorboat from the fishing villages along the shore. The steamer *O. K. Service* brought a hundred relatives and friends of the fishermen who were lost from the LaHave Islands and the western part of the county.

The procession formed at the courthouse, and at half past three made its way down the steep streets to the waterfront, where businesses were flying the flag at half-mast. The First Battalion Band, which led the parade, was followed by sea captains, merchants, robed choirs of the various churches, officers of the Canadian government ships *Arleux* and *Arras* in port for the occasion, town council members, as well as clergy from the town and from neighbouring communities. The crowd marched to the wharves of Zwicker & Company, where a service was held in memory of "those brave sons of Lunenburg County, who lost their lives on the great waters and in prosecuting the fishing industry."

With people crowding the wharves and the decks of schooners, with the band and the choirs now on board a newly built vessel at the head of one of the finger wharves, Mayor Arthur W. Schwartz conducted the proceedings that opened with a well-known hymn:

O Holy Spirit, Who dids't brood,
Upon the waters dark and rude,
And bid their angry tumult cease,
And give for wild confusion, peace;
O hear us when we cry to Thee
For those in peril on the sea.

Ministers of the various town churches said prayers and read scripture. Rev. R. J. M. Park of St. Andrew's Presbyterian Church remembered seeing the vessels of the fleet gliding out of the harbour and disappearing "on the bosom of the Atlantic"; prayers had been said then, he recalled, for their safe return, and all looked bright until the storm off Sable Island carried the captains and the crews to their death. "Why death took our young, strong men, we cannot understand, but bow our heads to God's will."

After the hymn "Lead Kindly Light," Rev. W. E. Ryder of St. John's Anglican Church read the names of the men who did not return: Lunenburg County names like Captain John D. Mosher, Captain Charles Corkum, Warren Wagner, Parker Wamback, Basil Shankle, Ladonia Whynacht, Caleb Baker, Rounsfell Greek, Hastings Himmelman, James Tanner.

Following the committal a large floral anchor and wreaths were cast on the waters by Mayor Schwartz. A volley was fired from the government ships in the port, the band played a march of the dead, the hymn "Jesus, Lover of My Soul," and Canon Harris, Anglican rector of Mahone Bay, offered a benediction.

The *Lunenburg Argus* described the service as impressive throughout. The ceremony that mourned "the finest and best in the county," declared the *Lunenburg Progress-Enterprise* on October 6, was the most touching and soul stirring ever witnessed in the town.

Unbelievably, sadly, the ceremony was repeated a year later, for a terrific storm on August 24, 1927, took many more lives at Sable Island. This time four schooners of the Lunenburg fishing fleet sank beneath the tumultuous waves—the *Joyce M. Smith*, the *Mahala*, the *Clayton W. Walters*, and the *Uda R. Corkum*. Over eighty more men were lost off the "graveyard of the Atlantic."

Once again, on Sunday, October 9, thousands of people poured into town, by car and by boat, to pay their respects to the fishermen who had left their homes that summer to win their livelihood and failed to return.

Never in its history had the County of Lunenburg known such widespread sorrow, declared the local press. Flags all over town flew at half-mast. The *Progress-Enterprise* commented that the previous year's disaster was seen at the time as the greatest tragedy that could befall the hamlets and villages of Lunenburg County; but, now, the loss of four vessels and their crews led to "a sadness beyond the depth of human ken."

A service was held in the square surrounding the bandstand, a fitting place, it was said, for there stood the monument erected in memory of the gallant young men of the town who had made the supreme sacrifice in the Great War. Once more the ceremony was presided over by Mayor Schwartz. The battalion band and the massed choirs of the town led the singing of hymns.

> Jesus, Saviour, pilot me
> Over life's tempestuous sea;
> Unknown waves before me roll,
> Hiding rock and treacherous shoal;
> Chart and compass come from Thee,
> Jesus, Saviour, pilot me!

Scripture was read and a male quartet sang the moving and appropriate words of Tennyson's "Crossing the Bar."

> For tho' from out our bourne of Time and Place
> The flood may bear me far,
> I hope to see my Pilot face to face
> When I have crost the bar.

Following the reading of the Roll of the Dead, once again the procession made its way down the hill to Zwicker's wharf: the band, followed by the choir, captains and crews of the fishing vessels, officers and crews of the government ships, heads of the outfitters and fish firms, the clergy, the mayor and town council, and the lieutenant-governor, James Cranswick Tory. Wreaths were conveyed to the waterfront on a draped fish barrow carried by four Boy Scouts. The parade was followed by "a great concourse of people."

Once more wreaths and flowers were scattered on the waters of the harbour. Relatives of the crew members of the *Clayton W. Walters* from Vogler's Cove and Port Medway brought "a profusion of beautiful flowers, wreaths, crosses and cut flowers" for the solemn occasion. A particularly moving moment came when a wreath from the outfitter Acadian Supplies, in memory of the crew of the *Uda R. Corkum*, was deposited on the water by Uda, the young daughter of Captain Freeman Corkum; the captain had had the vessel built in 1918, named it after his daughter, had sailed her as master for many years, and remained one of her owners.

From the bridge of the CGS *Arras* the choirs led by the band sang the hymn "Safe in the Arms of Jesus." A salute from a firing squad on the *Arras* concluded the service.

In the coolness of a typical autumn day, one journalist wrote, and on "the rippling waters of the sheltered harbor, where the fishing fleet is now riding safely at anchor, the wreaths and flowers were scattered in memory of the brave heroes of the deep, who return no more to home and loved ones."

In two seasons, the August gales had claimed the lives of more than 130 men in the Lunenburg fleet.

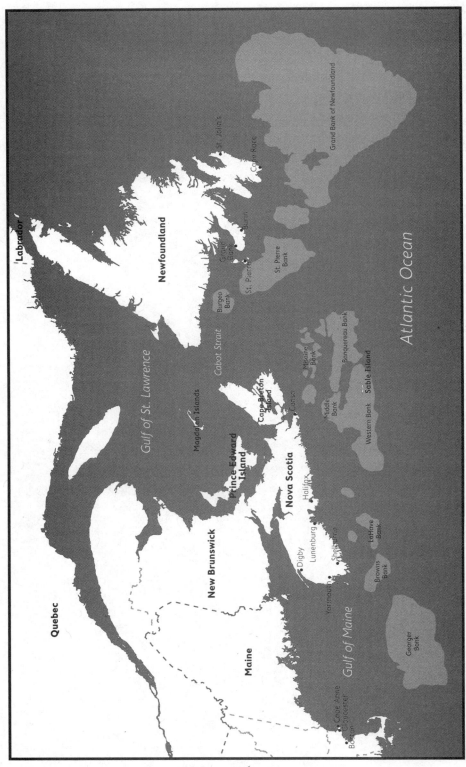

LUNENBURG
and the BANKS FISHERY

I T ALL HAD TO DO WITH FISH.

There existed in the western North Atlantic a triangle—a piscatorial triangle—based on the pursuit of cod. A curious triangle perhaps, for on a map it appears to be more like a straight line. But let us imagine a triangle anyway, with its three sides and three vertices. In the north, Newfoundland, closest to the fishing grounds of the Grand Banks, where cod were so ubiquitous they were simply called fish. The southern point is located firmly in Gloucester, the Massachusetts town that became the fishing capital of New England; here the nearest fishing grounds were located on Georges Bank, but Gloucestermen had been fishing all over the banks since the early eighteenth century. And the final vertex of our imaginary triangle is Lunenburg, the town on Nova Scotia's South Shore that over time became the principal banks fishing port in Atlantic Canada.

Three different fishing communities, three different countries, but in their pursuit of fish on the banks they would have much in common, including the terrors of the North Atlantic storms.

> Why you go to Lunenburg an' get up on one end o' the wharves there an' look down o'er, well it was the same as woods. Spars looked like a forest, you. The harbours layed full o' vessels... fishin' vessels.... It was somethin' pretty to see. And the vessels would leave for out in the spring o' the year...after the men that went had their plantin' done. Got the gardens all in before they went down on the trip, you....I can see them big white sails yet....Yes it was a pretty sight.
>
> —From Peter Barss, *A Portrait of Lunenburg County*

By the 1920s, when tragedy struck two years in a row off Sable Island, Lunenburg had earned its reputation as the major deep-sea fishing port on Canada's east coast. Indeed, the *Halifax Herald*, in a review of the year past on December 31, 1920, claimed that Lunenburg, with its 120 sailing vessels, many equipped with gasoline auxiliary power, had "without a doubt, the finest fleet of sailing vessels of any single port in the world." And the town was portrayed in the monthly fisheries trade journal *Canadian Fisherman*, in August 1926, as "the haven of the largest fishing fleet in North America." Writing about her husband's native land in *Bluenose: A Portrait of Nova Scotia*, Dorothy Duncan, wife of novelist Hugh MacLennan, praised Lunenburg as a seaport with "a fame that rides the seven seas," its men "famous around the world as the finest sailors afloat." It's one of history's small ironies that these men, who became such successful fishermen, sailors, and shipbuilders, had begun life in Nova Scotia as farmers.

Although the British had conquered Acadia in 1710, it was not until 1749 that the capital was established at Halifax, only four years before the founding of Lunenburg. The fortifications at Halifax were intended to counter the influence of the formidable French fortress at Louisbourg on Cape Breton Island. There was concern, as well, about the loyalty of the large French and Catholic Acadian population scattered about the colony. With little hope of finding suitable emigrants at home, the British authorities decided to search farther afield for solid Protestant stock. Public notices were posted in towns in central Europe promising land, free rent and no taxes for a period of time, and provisions to help settlers get established. The healthy climate and the productive, fertile soil of the province were extolled. Thus enticed, people left their homes in southwestern Germany and in nearby Montbéliard, a French-speaking principality to the north of Switzerland, and crossed the ocean to the New World. Although other trades were represented, such as masons and carpenters, shoemakers and tailors, butchers and bakers, miners, schoolmasters, and brewers, the largest proportion of newcomers by far gave their occupation as farmers.

The 1,453 "foreign Protestants," the founders of Lunenburg, after some time in Halifax, came ashore at Rous's Brook early in June 1753. Though mainly German, there were Swiss and French among them. Lands were granted and eventually cleared. Agricultural tools and other items needed for farming were given out, as were lumber and nails for the building of houses. According to Mather Byles DesBrisay's *History of the County of Lunenburg* (a book which Marq de Villiers described as "idiosyncratic, encyclopedic and occasionally erratic all at once"), the government in 1754 sent to the inhabitants "74 cows, 967 sheep, 114 pigs, and 164 goats, besides poultry." Starvation was averted in the winter of 1755 when rations of bread and beef were issued; many cattle died of distemper. As late as 1759 the people were still drawing rations of one pound of flour a day.

Colonel Charles Lawrence, who had been put in charge of the operation, felt the settlers needed a tight hand to begin with. In a letter to the governor he declared that he found them to be "inconceivably turbulent, I might have said mutinous, and are only to be managed like a great ship, in a violent storm, with infinite care, vigilance & attention." Lawrence was, after all, a military officer, and used to being in charge. And the settlers—who had faced a long and difficult Atlantic crossing, had waited in Halifax for many months, and then were not immediately let off the ships in the harbour—perhaps had reasons to be a bit grumpy. It didn't help that there was heavy rain in the first weeks. Even as the settlement evolved, Lawrence continued to complain of "ye Sloth of these Germans, & their unwillingness to do any part of what they call ye King's work." They expected to be paid for their labour in public works, he claimed, and they even ran off with boards needed for building the blockhouses for their own protection. Nevertheless, Lawrence was hopeful, concluding that the more he looked around, the more promising the prospects were. "We have fine land, fine timber, fine fish, & in great abundance. People so happily situated," he lamented, "ought to be both grateful and industrious."

When the settlers came ashore they found about 150 hectares of land already cleared. There had been French farms in the area of what was then known as Merligueche since about 1630, after the French explorer Isaac de Razilly had established a fort at nearby LaHave. By 1753, however, only one Acadian household remained, that of a man known as "Old Labrador," who was likely part French and part Aboriginal. The Acadians ceased to be a serious concern only two years later, in 1755, when, in the infamous *grand dérangement*, some ten thousand people were deported from the province.

Although any danger the French represented was largely removed by the expulsion of the Acadians, the Mi'kmaq were perceived to be a greater threat to the new settlement. Colonel Lawrence was naturally much concerned to build strong defences. Winthrop Pickard Bell, in his history of the foreign Protestants, claimed that "Indian barbarities" in these early years blighted what small farming settlement had begun and prevented further efforts. There were incidents in the countryside, to be sure, and some of them were bloody—most notably, an attack on the Payzant family on an island in Mahone Bay in 1756, and the Ochs family in the Northwest Range, who were "scalped and carried off" in March 1758. On

Lunenburg's natal day, June 7, 1886, the Rev. R. C. Caswall, rector of St. John's Anglican Church—with considerable hyperbole not to mention a fair bit of melodrama—chronicled poetically the perils and perfidy that had faced the early settlers in Lunenburg, "those worthies so brave":

> Who landed right here in the midst of red savages,
> And were often exposed to their murderous ravages;
> ...
> Not fearing the tomahawks, axes and knives,
> The spears and the arrows, the bullets of lead,
> Which assaulted them waking, or slew them in bed...

The newcomers had arrived, after all, in the middle of a long-standing war between the British and the Mi'kmaq, who were often backed by their French allies. There had been treaties made and treaties broken. In May 1756 Lawrence's "scalping proclamation" awarded £30 for every male Mi'kmaq prisoner over age sixteen, or £25 for his scalp, and £25 for every woman and child brought in alive. As Daniel N. Paul wrote about his people in *We Were Not the Savages*, the Mi'kmaq paid a heavy price attempting to resist the seizure of their lands. By the early 1760s hostilities had pretty much ceased, and the town of Lunenburg had not itself been attacked. The Mi'kmaq people were descending into poverty and destitution. According to Judge DesBrisay, a happy period had dawned, when every man, instead of "the wild whoop of the Indian," could listen to the sweeter sounds of his wife and children calling him from toil.

It was, then, something of a bumpy start, but the town flourished, as did the county around it. Houses were built, gardens planted. A successful farming community slowly developed. By 1827 it was reported that the settlers were growing wheat and other grains, potatoes, cabbages, apples, and hay; they were raising horses, cattle, sheep, and pigs. That the agricultural settlement experienced success even earlier can be seen in an 1818 letter concerning an attempt to set up a local agricultural society. Lunenburg, it was thought, could give practical lessons in the art of husbandry, for it had long been engaged in the pursuit. "That industrious settlement is a model to all others: by a careful cultivation of the soil, and by attending to the products of the garden, it has amassed more real opulence, and has a greater command of cash, than any other county, and is a public example of what can be achieved in Nova Scotia by the plough and the spade."

Because the foreign Protestants came from inland districts in central Europe, they had no experience in the fisheries. Only two of the original settlers claimed to be "mariners." In his report on the settlement in 1761, surveyor general Charles Morris observed that they spent most of their time clearing and improving their lands: "They have no inclination for the fishery, tho' well situated for that purpose." Despite the fine harbour and ready access to the sea, early activity was directed towards farming. As Harry Hewitt, a former citizen of Lunenburg, remarked in his history of the town, the settlers were brought out as "agriculturalists," primarily to carry on farming, and "they did not take kindly to fishing." Bell declared this lack of interest in the fisheries to be "a mildly amazing thing," noting "their advantageous location for that industry."

As early as July 2, 1753, Colonel Lawrence had seen the possibilities of the location, even foreseeing the direction in which Lunenburg's destiny lay: "Ship-timber, boat-building, barrel-staves, & hoops, boards & plank & ye fishery, are all articles by which these people, were they conversant with business, might acquire…an immediate livelihood…." If thus employed in the winter, they could make money and purchase stock for the spring, and would soon become both farmers and traders. Alas, Lawrence thought this grand future could only come about if English settlers were allowed into the community.

The Lords of Trade and Plantations, behind the project in London and anxious to be rid of the burden, urged the governor "to induce the settlers to follow the fisheries to a greater extent." The number of vessels did increase slowly over the years, supplying fish as food as well as fish oil and fertilizer, thus supplementing the community's agricultural endeavours. Hewitt claimed that fish oil used as an "illuminant" was more important than fish as food. According to B. A. Balcom in *History of the Lunenburg Fishing Industry*—the most comprehensive study of the topic—there were six vessels fishing out of Lunenburg by 1767. A couple of men would head out from shore in a small boat and bring back fish to be pickled or dried on land. This was the beginning of the inshore fishery, and for some time to come it would be the most important branch of the county's fisheries. Even so, Balcom noted, other communities further down the shore, such as Liverpool and Shelburne, built their early economies around the fishery, while Lunenburg, blessed with more fertile soil, for the time being stuck mainly to farming.

Eventually, however, Lunenburgers ventured farther and farther out to sea. In time, a small fleet of coastal boats began trading with Halifax, carrying cordwood, lumber, and agricultural produce up the coast to the capital. Ruth Fulton Grant, in *The Canadian Atlantic Fishery*, claimed that by 1795 hundreds of men with small farms were absent in the summer fishing, leaving the women behind to hoe the potatoes and to look after the farm work generally. The combining of fishing and farming, or at least some gardening, thanks to female labour at home, would become a Lunenburg custom.

By the early 1800s Lunenburg vessels were venturing as far as the coast of Labrador in search of cod. Balcom wrote that by 1829 Lunenburg had a fishery consisting of two brigs, sixteen schooners, and thirteen shallops fishing off Labrador, using caplin as bait.

When loaded the vessels would sail for Lunenburg, Cape Breton, or Newfoundland, where the fish were removed from the hold, washed thoroughly, and spread on the beach to be "made" by the sun. After a couple of weeks, they were put back on the vessel and taken to Halifax to be sold. The *Bridgewater Bulletin* of January 3, 1928, observed that the Labrador fishery came into existence soon after the colonization of the town, the fleet of a few small vessels leaving the home port about the first of June and returning in late August. "In the pioneer days, stern necessity demanded that the settlers, who were tillers of the soil rather than fishermen, should procure their subsistence in various ways and fishing was early recognized and entered into by a large proportion of the male population as a productive source of income, until with each succeeding generation, following in the footsteps of ancestors, there has been built up an industry which has become the very backbone of the town and county."

Deep-sea fishing took place as well on the fishing grounds, or banks, closer to home off the Nova Scotia coast. The men fished for cod with hooks and lines (handlines) from the decks of vessels, and the bait used was herring and mackerel that had been obtained earlier on shore. Cleaned, split, and preserved in salt in the hold, the cod was dried upon return to port. These ventures were the first small steps towards the banks fishery.

In these early days Lunenburg was also establishing important trade links abroad. Lumber, wood products, fish oil, dried and pickled fish, and vegetables were taken to the West Indies, the vessels returning with spirits, molasses, sugar, and coffee. Because Catholics were required to eat fish on Fridays and during Lent and on other holy days there was also a demand for fish in southern Europe and elsewhere; however, the trade connections with the Caribbean were especially important and would remain so throughout the fleet's history.

The county surrounding Lunenburg was also being populated during these years. From the earliest days some of the townspeople moved onto farms in the countryside or established small fishing communities along the shoreline. Several of the founding families settled at Blue Rocks and Stonehurst (often known locally as Black Rocks), farther out to sea towards the end of the Lunenburg peninsula. The sheltered coves along the rocky coastline were within rowing or sailing distance of the nearby fishing grounds. Gregory P. Pritchard, who wrote about Blue Rocks, observed that the sea provided a livelihood, but one that was seasonal, and that cyclical incomes fostered frugality and insecurity. Succeeding generations, he said, would know nothing but fishing. Boys grew up in boats and the sea became second nature to them. Many were engaged in shore fishing, but others would find berths on the schooners sailing out of Lunenburg.

Settlement also spread along the LaHave River, southwest of Lunenburg. Joseph Pernette, a retired military officer who had come to Nova Scotia with the foreign Protestants, was granted a tract of land on the west side of the river in June 1765. Many families of Germans and others settled at Getson's Cove, which later became the village of LaHave. Farmland was also granted early on around Summerside (later known as Dayspring) and elsewhere along the river. Vessels from the small coves along the river and among the islands at the mouth of the river took part in the inshore and offshore fisheries.

Over time Lunenburg had developed a fairly mixed economy, and the men of the county, like Nova Scotians elsewhere, could put their hands to many tasks. By the middle of the nineteenth century judge and author Thomas Chandler Haliburton could write: "The Nova Scotian…is often found superintending the cultivation of a farm and building a vessel at the same time; and is not only able to catch and cure a cargo of fish but to find his way with it to the West Indies or the Mediterranean; he is a man of all work but expert in none."[1]

※ ※ ※

The year 1873 marked a turning point in the fortunes of the Lunenburg fishing industry. Five schooners, equipped with trawls and dories, sailed from the port that year to the Western Bank off Sable Island. The fishing didn't go all that well, and four of the skippers abandoned the enterprise and headed once again for Labrador. Captain Benjamin Anderson, however, in his schooner *Dielytris*, remained on the banks and made a catch of 1,850 quintals for the season (a quintal, used as a measure for dried salt cod, is a hundredweight or 112 pounds). This was considerably better than could have been expected from the Labrador fishery with the same investment in equipment and men, so Anderson returned the following year, this time accompanied by four other vessels. Although some seasons were hampered by a shortage of bait, the offshore banks fishing fleet continued to grow. In 1877, Balcom noted, only five years after Anderson's first voyage, there were somewhere between thirty and forty vessels from Lunenburg County trawling the banks.

Fishing on the banks was not new, of course, for the North Atlantic fishery was at least a couple of centuries old by the time Lunenburg was founded. John Cabot, the Venetian who crossed the ocean in 1497 on behalf of Henry VII of England, supposedly famously reported that the sea was "swarming with fish, which can be taken not only with the net, but in baskets let down with a stone." It's likely that European fisherman were in the area before Cabot's voyage, but the first recorded cargo of cod from the fishing grounds off Newfoundland arrived in Bristol on the *Gabriel* in 1502. French, Basque, and Portuguese fishermen were also heading annually to the banks by this time. In 1617 Samuel de Champlain reckoned that between eight hundred and one thousand French vessels were engaged in the North American cod fisheries every year.[2] Over three hundred years later the French remained a strong presence on the fishing grounds: in March 1927 fishermen in Saint-Malo paraded from the cathedral to the wharves, carrying on their shoulders the fishing dories for ninety schooners about to sail for the banks off Newfoundland. As anthropologist Peter E. Pope declared, John Cabot may have been seeking Japan, but his greatest discovery was cod.

The banks are a series of shallow submerged plateaus on the coastal shelf of northeastern North America, extending from the Grand Banks off Newfoundland in the north to Georges Bank off Massachusetts in the south. The abundance of life there is caused by a mingling of waters: the Labrador Current carries cold water, cool air, and nutrients from the north to meet the warm waters of the Gulf Stream coming up from the south. As well, the St. Lawrence Current brings more nutrients in the runoff from the continent. This rich environment produced cod by the millions but also the fish on which the cod fed: mainly squid and caplin, as well as tiny sea creatures that nourish innumerable kinds of marine life. Unfortunately, the conditions also produced dense and persistent fog, and sometimes large ice floes and dangerous icebergs. In exploiting this resource, the continental Europeans salted the fish aboard and took them back home. The British, at a disadvantage because their supply of salt was limited, tended to dry the fish on shore. The British colonists in New England were also fishing on the banks long before the Lunenburgers.

James Eisenhauer, the "merchant-prince of Lunenburg" who would make a large fortune in the West Indies trade, claimed that Lunenburg only gradually began to specialize in the banks fishery. He maintained in 1877 that Lunenburg fishermen usually made two trips a year for cod: to the Western and LaHave Banks, Banquereau (Quero), and sometimes the Grand Banks in the spring, but also to the Gulf of St. Lawrence in the summer. With the introduction of trawling from dories in the 1870s, fishing on the banks steadily increased, according to Balcom, and the Labrador fishery declined.

As well as the improved technology at sea, there were important changes ashore. The men behind the banks fishery, entrepreneurs like James Eisenhauer, Lewis Anderson, and W. N. Zwicker, formed trading companies so successful that by 1878 there were eighteen vessels sailing to the West Indies and some to American ports. They had demonstrated the usefulness of the trawl and provided the capital necessary for the new vessels and gear. Their outfitting firms, such as W. C. Smith & Company, Acadian Supplies, Zwicker & Company, Adams & Knickle, and the Lunenburg Outfitting Company, grew, with their own wharves, warehouses, and fleets of vessels. The size of the vessels gradually increased after the fleet switched to the banks, and the number of fish they caught eventually overtook the catch of the many boats engaged in the inshore fishery. By the late 1880s Lunenburg had assumed a dominant position in the cod fishery of the province. As Balcom put it, these entrepreneurs "launched Lunenburg in the classical era of the bank schooner fishery."

The fishing industry also flourished elsewhere in the county, most notably at the mouth of the LaHave River. Fishing vessels and larger trading vessels were built locally. Schooners returned from the banks with cargoes of fish to be dried on flakes along the shores of the river. Ships' chandlers operated in the larger communities downriver such as LaHave and Ritcey's Cove (now Riverport). Before leaving port, the vessels had to be equipped with "everything from an extra spar to a sail needle, salt, fuel for cooking, food and water, and, no doubt, a keg of rum for the crew," obtained from outfitters such as the Himmelman Supply Company or the LaHave Outfitting Company. As in Lunenburg itself, we're told in *Historic LaHave River Valley*, large houses were built along the river by prosperous sea captains and successful merchants.

There were of course developments in the fishery elsewhere in the province. Halifax, for instance, according to Ruth Fulton Grant, had a fleet of fishing schooners, but its future lay in distributing fish from communities along the coast as well as from the Gaspé and Newfoundland. Shelburne, Yarmouth, and Digby Counties had been populated chiefly by Acadians and then by Planters and Loyalists from New England, and their history and proximity to the "Boston states" influenced their development. Yarmouth, for instance, became prominent in shipbuilding and shipping. The town of Liverpool, in Queens County next to Lunenburg, was founded by New Englanders such as Simeon Perkins before the American Revolution, and many of the settlers there brought schooners and fishing skills with them.

Lunenburg, on the other hand, had fewer direct ties to New England and developed in its own fashion, perhaps partly because of its different, mainly German, ancestry. Its prosperity in the 1880s and 1890s was based largely on what became known as the "Lunenburg cure," a dried fish that kept well and was relatively cheap, a low-grade product that found a ready market in the West Indies. Originally the port traded almost exclusively with the British colonies, but by the later 1890s more of the fish was going to Cuba and Puerto Rico.

The key to Lunenburg's success as a fishing port seems to have been the all-inclusive nature of its fishery. As Balcom observed, "the county's vessels could be built, outfitted, provisioned, manned, and even insured locally." Shipyards throughout the county built schooners and dories, and men in lumber camps and mills produced the timbers needed for construction. Farmers provided produce. Businesses from coopers to ships' chandlers flourished along with the fishery. Harry Hewitt outlined the intricacies of the local industry: "To prosecute the fisheries, vessels and boats are needed. The forests supply the wood, the artificers fashion it, the painters protect it, the caulkers make the seams between the planks watertight. The sailmakers fashion the canvas, the blockmakers turn the lignum vitae into hoisting devices, the riggers support the towering masts with ropes. The blacksmith creates a hundred articles intended to fasten and to hold. The outfitters supply everything necessary to maintain life in comfort in a home away from home. The foundry equips the galley with a range and the hold with a gas engine. Thus a far-flung community finds its livelihood provided by filling the needs of the fishing fleet."

Early in the twentieth century Robert R. McLeod, in *Markland or Nova Scotia*, spoke of Lunenburg's "gigantic" fishing industry which exceeded in value that of any port in the Dominion; it also outstripped, he said, "in number of vessels engaged in deep-sea fishing the once famous fleet of the New England States, until today Lunenburg is justly entitled 'The Gloucester of Canada.'" The people of Lunenburg, he suggested, in the language and sentiments of his time, had finally moved forward into the modern world: "The latent energy of the old stock has been aroused by the spirit of the age, and this quaint old German outpost, that dozed and dreamed amid vacated blockhouses and cabbage yards for a century, awoke to its opportunities, and Lunenburg is full of life, and thrift, and hope, and beautiful for situation, as it overlooks the restless view of ocean, and islands, and headlands that fade into the dim perspective of distance."

Fishing naturally varied in different parts of the province. An interesting view of the Lunenburg operation, reported in Frederick William Wallace's autobiography *Roving Fisherman*, was expressed, with more than a hint of scorn, by a winter haddock fisherman in Yarmouth in 1912: "All salt-bankers. Fish only in summertime—March to September—and then lay up for the winter, or go freighting to the West Indies carrying dried fish, lumber and shingles to Barbados, Porta [*sic*] Rico or Cuba. They go to Middle Bank, Quero and Grand, let go the anchor and set the dories out. They're codfishermen—they don't want haddock, halibut or anything else but cod. That's their business. If a Lunenburger ever was to go winter haddocking, and making flying sets, he'd probably lose all his gang, all his dories and all his gear."

That's not a bad description of the Lunenburg fishing industry, but it does need a bit of fleshing out.

To begin, the fishermen sailed to the banks on schooners, among the most beautiful wooden sailing ships ever built. Typically the banks schooners had two masts, the mainmast taller than the foremast, and were rigged fore and aft along the length of the vessel. The sails soared above the deck and the trim, usually black, wooden hull like giant white clouds. "No other sailing boat," declared Peter Carnahan in *Schooner Master*, his book about Lunenburg County boat builder David Stevens, "has lines quite as balletic, as reminiscent of dancers, racehorses, whippets—that special category of living things that triumph over our normal gravity and inertia. No other sailing boat has her long sweep of hull echoed and amplified by the series of fore-and-aft rigged sails lying close to it, repeating in the many curves of the canvas the curves of its wood. With all her sails and topsails set, her shoulder to the wind on a close reach, she has the look not just of a gull, but of a whole company of gulls riding high on the wind." David Stevens once said of a schooner: "She's the closest thing to a human being that a man can build."

Raoul Andersen, in *Voyage to the Grand Banks*, put it more prosaically, "A banks schooner flying under sail is a lovely picture. It is also the fisher's home at sea and a machine, a tool, for catching fish." Though splendid to behold, the schooners were above all meant to be useful. As the fishing industry progressed, the design of schooners improved and their size increased. In the 1850s, at the dawn of the golden age of the great banks schooners, the average vessel measured about 45.4 tonnes, and by the early 1900s the size had become 86 to 91 tonnes[3]—the *Bluenose*, launched in 1921, was 89.8 tonnes.

The schooners were skippered by men with a great deal of solid experience, though they usually had little formal education. The editor of *Canadian Fisherman*, Frederick William Wallace, who knew

"...the soaring, mind-boggling beauty of the schooners."
—the Lunenburg fishing vessel *Bernice Zinck*

many in his time, praised the "marvellous seamanship and uncanny intuition" of the old-time captains, "resolute men" who year after year took their ships out to the banks or around the world. They had little equipment with which to navigate, mainly a compass and a deep-sea lead to be dropped overboard to find the depth of the water (in his novel *Blue Water*, Wallace called the lead "the fisherman's third eye"). A sextant could be used to fix the ship's position, but that required taking a sighting from the sun or the stars and was useless in fog or cloudy skies. In heavy weather the captain could navigate by "dead reckoning," using the vessel's speed and direction to attempt to estimate his position. It's important to remember that the men who sailed the schooners, the crew as well as the captain, were skilled sailors as well as fishermen.

The Lunenburg fishing season lasted about six months, extending from around the middle of March until the end of September. In general, three trips were made to the banks. The short first trip in the spring, introduced in the early 1920s, was called the "frozen bait" trip because the bait used was herring or mackerel that had been preserved frozen over the winter; it was obtained from ice houses in such places as Sydney, Canso, or Queensport on Chedabucto Bay, and frozen herring was also picked up in the Bay of Islands, Newfoundland. Once back in port and unloaded, the vessels were refitted for the spring trip, which ended in early June. The third, or summer, trip followed almost immediately, returning late in September, the men being away sometimes for almost four months. During the winter most of the Lunenburg schooners lay at anchor in the harbour, though some were fitted out for coasting and carried fish to the West Indies, returning with cargoes of salt for use during the next fishing season.

Except for the first trip of the year, a number of the vessels usually went to the Grand Banks off Newfoundland. Caplin and later squid were used for bait during the summer. For those schooners out on the farthest banks, it was necessary to run in to Newfoundland ports such as Cape Broyle or Aquaforte every few weeks for supplies including water, bait, and ice. Bait only lasted two or three weeks, and drinking water could become rather green and slimy. Since schooners from Newfoundland and New England were also looking for bait, much time could be wasted searching for a port with a sufficient supply. Later on, cold storage facilities would help to fix the bait problem.

The main catch of the Lunenburg fleet was, of course, cod, *Gadus morhua*. They were, above all, plentiful, and their flesh dried quickly and preserved well. As a cheap source of protein, they became a staple food for the poor. It's not surprising that cod also figures largely in the local cuisine, from Newfoundland's "cod au gratin" to a specialty in Lunenburg called "housebanken"; indeed, the first recipe in the popular *Dutch Oven* cookbook, put out by the Ladies Auxiliary of Fishermen's Memorial Hospital in 1953 and reprinted many times, is "Dutch Mess or House Bankin," requiring "1/2 lb salt dry cod, 6 large potatoes, 2 oz salt pork, Butter size of egg, 2 onions, 1 tbsp. vinegar, 2 or 3 tbsp. cream."

Dory fishing on the vast Atlantic.

Cod and the fishing industry were as important in the economy of Atlantic Canada as the beaver and the fur trade were in the rest of the country. As Harold Innis observed in *The Cod Fisheries*, the St. Lawrence River facilitated westward expansion and concentration on fur, lumber, pulp and paper, minerals, and wheat, whereas the bays and harbours of the East Coast led to expansion eastward and a concentration on one resource: fish. Innis also noted that in the fishing industry, since the ship was the largest technical unit and the initiative of the individual fisherman was of paramount importance, large centralized organizations, such as those in the fur trade, did not develop.

Once anchored on the banks, all sails furled except for a triangular riding-sail hoisted on the mainmast to keep the vessel steady, the men fanned out from the schooners to fish from dories. Small wooden boats with a flat bottom, very sturdy and seaworthy, dories were painted a yellowish orange or "dory buff" colour to make them as visible as possible in fog, snow, or at dusk. "It needs be bitter weather when the dories cannot go," said a line in a poem in the *Canadian Fisherman* of March 1914. There was nothing much aboard to save a doryman in trouble, however, as historian M. Brook Taylor noted in *A Camera on the Banks*: safety equipment was "generally limited to a good supply of food and water, a sail, and a fog trumpet."[4] Because of their removable thwarts, the dories could be stacked on top of one another, or "nested," on the schooner's deck, with the plugs removed for drainage.

The dory in its time was of course an innovation. As Newfoundlander Otto Kelland remarked in *Dories and Dorymen*, while the owners of schooners and fishing skippers became well off with the advent of new ways of fishing, ordinary fishermen were in some ways worse off, for "it took them from the relative safety of a vessel's deck out amongst towering seas, fog, ice floes and storms with one and one-eighth inch of a dory's bottom between them and the great beyond."

In the early days, when the Lunenburg fleet headed mainly for Labrador, fish had been caught from small boats by handlining—fishing by holding in one's hands single lines with baited hooks. Later on, the same method was used to fish from schooners. In the words of Gloucester historian Joseph Garland, "a dozen or so crew lined the rail, tending a brace of handlines each, hoping for the wily codfish or the hungry halibut to convene for dinner twenty fathoms below the vessel's bottom." Handlining from dories began in the 1860s. One man fished alone in his dory, operating as many lines as he could handle, usually two but sometimes more. Handliners used less bait than trawling vessels, Balcom maintained, and occasionally a good and fortunate crew could become "highliner" of the fleet—the schooner having the largest catch of the season. Some fishermen believed that bigger and better fish were caught by the handlining method. Although the Lunenburg fishermen turned mainly to trawling, the nearby port of LaHave was particularly noted for continuing to employ handlines.

Trawl fishing (or longlining), a method that was used increasingly as the fishery took to the banks in the 1870s, was more complicated, led to greater costs in outfitting and bait, but caught more fish more easily. (The term "trawling" here should not be confused with the later, modern trawlers that catch fish by dragging cone-shaped, or bag-like, nets over the bottom.) The trawl lines consisted of two parts. The main or ground line, made of heavy tarred cotton, was stretched out a kilometre or more from the schooner by the men in dories. Each dory would take a different direction, the vessel thus forming a hub from which the trawls radiated like the spokes of a big wheel. Attached to the ground line every metre or so were ganglings, about half a metre long, also made of tarred cotton, with a single hook attached. The hooks were baited before the men left the vessel; the trawls, each of which could have as many as eighteen hundred hooks, were coiled in wooden tubs and taken aboard the dories to be paid out as the dory pulled away from the schooner. Each end of the trawl was anchored and marked by a flagged buoy. After giving the fish time to bite, the men would return to the trawls and bring the line to the surface by a method known as "underrunning": with the ground line still connected to the flags at either end, the man in the bow would haul up the trawl and remove whatever fish had been caught; the other man would rebait the hooks as they came to him and pass the freshly baited end of the trawl over again into the sea, on the opposite side from which it was coming aboard. In fine weather the men would return to the schooner, unload the fish, and return to the trawl again, as often as four times a day.

The dories were lowered from the deck for the first set at about four o'clock in the morning, often before breakfast. When they were finally hauled aboard after a long day's work, the dorymen would still have to help the crew on board get the cleaned and dressed fish into the hold. If the fishing was good, it could be well past midnight before they were finally able to turn in.

Catching fish by trawling was new technology, and as usual, wrote Balcom, it caused controversy. In a statement familiar to later generations, James S. Richard of Getson's Cove declared in 1877: "Trawling I consider very bad for the fishery, as the mother fish are taken and great quantities of spawn are destroyed. By hand-lining few mother fish are taken.... Canadians have been compelled to trawl in order to compete with the Americans."

The crew of a fishing schooner at the peak of the banks fishery was usually made up of about twenty men: the skipper, the cook, a salter, a throater and a header, fourteen fishermen divided into seven double dories if trawl fishing, and a flunkey. One of the men was second hand, or mate. The throater and header were usually young apprentices, fishermen in the making, and the flunkey was even younger. After the fish were landed from the dories, the throater would cut the throat of each fish, split it halfway down, and place it in a box-like device called the "kid." The header would then place the fish on the splitting table and remove the head and guts; at the same time, he would separate the liver for making fish oil. The fish was split to the tail and almost through the back and the backbone removed so that it would lie flat. With the splitting completed and the fish washed, the salter, down in the hold, would salt them heavily and pile them crosswise in bins called kenches. Light in the hold was provided by candles, sometimes using a "sticking tommy," a candle holder with a spike at one side and another at the base that could be stuck into a bulkhead or another wooden surface. At the end of the voyage, back in port, the fish were washed carefully to free them of slime, piled to allow the pickle to drain off, and finally dried in the open air on wooden stages or flakes by the people, often women, who cured them, called "fish makers."

Of the men aboard the schooner, the captain naturally stood above the rest. Loved or loathed, he was master of all he surveyed. He was responsible for the lives of the men, and had the power to drive both crew and schooner hard. He could order the men aloft in all weather, or send them out on the bowsprit to take in and stow the jib. It was his duty to find the best places to fish and he alone decided when the dories went overboard. Though he was likely not referring to fishing schooners, Charles Darwin's comment in *The Voyage of the Beagle* seems appropriate: "There is no such king as a sea-captain; he is greater even than a king or schoolmaster."

The other very important person on board was the cook. No food meant no fishing, a good cook led to a happier crew and better fishing. It was hot work, hard work, feeding so many hungry men, plus getting the supplies aboard in the first place. It could also be dangerous: almost always below deck in the forecastle, if the vessel foundered—rammed by a larger boat for instance—the cook was most likely to go down with it.

We can glimpse life at the other end of the scale through an exchange of letters by a family in the tiny community of Stonehurst,

LUNENBURG AND THE BANKS FISHERY

at the seaward end of the Lunenburg peninsula. The flunkey was very much at the bottom of the heap, but to many young boys it must have seemed a first step towards manhood. Elvin Tanner sailed aboard the *Elsie M. Hart* with his father Daniel in the summer of 1917 at the age of fourteen. Three years later his younger brother Victor joined his father and brother as flunkey aboard the *William C. Smith*. Elvin by this time had become a throater. Victor wrote home to his mother, and his sister Olive, in Stonehurst:

St. Lawerence   June 21, 1920
Dear mother I now wright you a few lions to let you now that we are all smart hoping you are the same. We diden get are bate not yet we expect to get are bate till the 23 day of June. The fish is plenty we expect to get a load of fish till the first day of august. The bate is very scares. we only got to dorys loas of bate. I have to close for this time. from you son Victor xx heres one kiss for the two xxxxxxxx

Willietta Tanner was naturally worried about the three members of her family aboard the schooner: "I hope Victor doesn't get so seasick. Look after him good Elvin. Do not let him on deck alone. Listen at papa and be good boys. I hope it won't be so stormy on the banks so papa doesn't have to work so hard." The practice of taking several family members as crew on the same vessel was common, and could have grave consequences.

Ralph Getson, curator of education at the Fisheries Museum of the Atlantic in Lunenburg, listed the multifarious duties of the flunkey: "There was one flunkey per vessel and his job was to help the cook, peel potatoes, set the table, empty the slops. He tended the painters as the dories came alongside. When dressing fish he kept the dress gangs supplied with fish. If he was keen he would cut out tongues, cheeks and sounds and salt them in a barrel. He would then sell them ashore when the trip was over. He was paid a pittance (if at all) and the tongues helped supplement his meagre wage. They sometimes had to fill the men's pipes while they were cleaning the fish and grease their leather boots. As one old fellow said 'Everyone was your boss...and your ass got kicked more than it got patted.'"[5]

The vessels in the banks fishery could be owned in various ways, but a co-operative system was popular in Lunenburg. Each schooner would be divided into sixty-four shares, which could be owned by the builders, the outfitters, the fishermen, or anyone in town who wished to invest. There could therefore be many share owners in a single vessel. The division of the spoils was complicated. From the total gross stock was deducted a commission of 2.5 per cent for the captain, the wages of the throater and the header, the cost of bait and ice and the curing of the fish that were landed. The remaining funds were divided equally between the owners and the crew. From the owners' portion was deducted another 2.5 per cent for the captain and the major costs of outfitting the vessel, including provisions, salt, and fishing gear. The cook's wages came out of the crew's share, as did several dollars per man for incidental expenses related to the operation of the vessel (for example, the cost of gasoline for an engine that hoisted the sails, in the unlikely event such a device existed on board). Apart from his 5 per cent, the captain also received one man's share. The amount received by each man fluctuated widely, depending on the quantity of fish caught, prices, and the expense of the voyage. The cook, the throater, and the header were, then, the paid employees aboard. The salter was entitled to one share. The flunkey, as noted, often didn't get paid at all.[6]

Handliners also divided the proceeds along co-operative lines. Shares among the fishermen could be divided equally, by count or by weight. "Equal shares" was the simplest procedure, but the crew members had to be known to be good workers. If the crew was less familiar, the captain might prefer to have each fisherman's fish, irrespective of size, actually counted as they came on board—a process that, as Mike Parker noted in *Historic Lunenburg*, could lead unscrupulous fishermen to throw away valuable larger fish in order to have more room for small fish.

Credit to prepare for the trip was obtained from one of the outfitting firms, and the outfitters were in turn financed by local banks. Each trip was settled as the cargo was sold. After the fish makers cured the catch, the owner or managing owner of the vessel would negotiate a price for the dried fish with the merchants, who also acted as outfitters. The merchants and outfitters often held ownership in several vessels to minimize potential loss. The price paid was determined by the market demand, so the fish might be held until the market improved, said Balcom, resulting in a delay of several months before the voyage's accounts were settled and the fishermen paid. The ship chandler could control the whole process: they outfitted the vessels in which they owned controlling shares and they were also in the business of buying and exporting the fish. As Mike Parker wrote, "There were as they say no poor outfitters or captains, only fishermen who often waited months to be paid. In the interim, they lived on credit from the company stores so that when it came time to settle accounts, little money actually changed hands."

The waning years of the nineteenth century proved to be "the highwater mark of the Lunenburg bank fishery." At the beginning of the twentieth century, according to Balcom, about two thousand men in 170 vessels were sailing each year to the banks to fish for cod. Routines had been established on shore to finance the fleet and to market the catch. But in fact, Balcom maintained, the fisheries appeared to be "tradition-bound," the spirit of inventiveness of earlier years dormant. Not everyone, however, thought Lunenburg's resistance to change was necessarily a bad thing. Writing in the

Canadian Fisherman of July 1914, Agnes G. McGuire observed that Lunenburg was still sending 125 vessels to the banks every year, all of them engaged in salt fishing. "With a strange conservatism," she wrote, "the Lunenburgers refuse to depart from the business they established over a century ago, but their conservatism has been one of progress in their particular industry and today the town with its splendid fleet of up-to-date, able schooners, its progressive business firms, and solid, substantial prosperity, can proudly claim the title of the 'Gloucester of Canada.'"

However, the times were changing. New technology in the form of gasoline engines was threatening the age of sail. Steamships were beginning to replace sailing ships. More markets were beginning to prefer fresh and frozen fish to dried fish. What's more, political changes were interfering with Lunenburg's traditional ways of doing things. The Spanish-American War of 1898 threatened the important trade links with the West Indies. Cuba declared independence and temporarily imposed heavy duties on Canadian goods including fish, while at the same time the Americans, who had taken over Puerto Rico, established protective tariffs in 1901. Fortunately, Lunenburg vessels were making record catches, and trade agreements later in the decade reopened the dried fish market. Lunenburg was helped too by the fact that its main competitor, Gloucester, the older and larger port in Massachusetts to which it was so often compared, was sending far fewer vessels to the banks. New Englanders were turning away from salt bank fishing, concentrating more and more on the fresh fish industry, developing refrigeration, and increasingly using steam trawlers. Lacking rail connections and cold storage facilities, Lunenburg was unable to handle fresh fish easily and so remained preoccupied with the traditional salt fishery.

At home, Lunenburg was losing business to Halifax. The capital, according to Balcom, was becoming the place to market fish, partly because it offered higher prices but also because fish were increasingly being exported on steamers out of Halifax rather than in

Lunenburg vessels. As well, Lunenburg captains and vessel owners were finding it more convenient to outfit their vessels in the city, for later in the season they would be selling their fish there. At the same time, Grant maintained, the Nova Scotia fishing industry in general was beginning to face something of a labour shortage. Coal mines, the iron and steel industry, railway construction, and other activities that needed labourers were attracting men away from the fisheries. Perhaps even more important, sons in families that for generations had been fishermen were increasingly attracted by the higher wages paid in the industrial centres of New England. Other young men were heading for the rapidly growing Canadian West. In the words of Harry Bruce, "As part of Nova Scotia's heritage, leaving home would long outlast the Age of Sail."

The early years of the new century were difficult times for the Lunenburg fishermen, a period when, it seemed, stagnation threatened and change might reluctantly be necessary. However, the fleet was about to achieve its greatest success during the First World War, which began in Europe in the autumn of 1914. Harry Hewitt claimed that the biggest and most prosperous year in the history of the fleet was 1917, when ninety-five vessels, with 1,884 men, landed a record catch of 266,215 quintals of fish, averaging 2,696 quintals per vessel, and fish sold that year at ten dollars a quintal. While the operating costs were high during the war, the profits were abnormally high, and there was an unusually strong demand for Lunenburg's dried fish. Competition from Europe was drying up: Norway in particular had been making inroads into the markets in the West Indies but due to wartime circumstances was selling almost all of its catch in Europe. It was, said Hewitt, halcyon days for the Lunenburg fishermen.

Fishing in wartime had its dangers, even though the fleet was far removed from major hostilities in Europe. German U-boats, it seemed, regarded fishing schooners as fair game. In September 1918 the *Canadian Fisherman* reported that nine Lunenburg County vessels, worth $264,000, with fish aboard valued at $136,000, had recently been sunk, as had a vessel from Yarmouth. In an item called "The Submarine Menace" the journal claimed that the practice of fishing from very seaworthy dories at least gave the fishermen on this side of the Atlantic a better chance of survival; British fishermen, many of whom were swamped in their overcrowded small yawls and lifeboats, were less likely to make it to land in rough waters.

A fisherman from the LaHave Islands, Kenny Reinhardt, had his own wartime story to tell. His schooner, its hold full of fish, was sunk by a submarine one dawn late in the war on the eastern part of the Grand Banks, over 250 kilometres from St. John's. The Germans raided the vessel for flour and other goods and some of the men's belongings before sinking it. The captain had been taken aboard the submarine and plied with liquor before being released. The men were in dories from Thursday evening until Sunday evening when they

were picked up by a Norwegian steamer, having rowed or sailed more than 160 kilometres.[7]

A curious and perhaps little known wartime venture involved the "mystery fleet." Three Lunenburg schooners, two from Digby and one from St. John's, were fitted out as Canadian fishing craft but with false nested dories concealing guns to confront German submarines. These "special service vessels" were under the direct authority of the commander-in-chief, North American and West Indian Squadron of the Royal Navy, and they were manned by naval ratings as crew. By late 1917 the decoy boats had become so well known to the enemy that most of them were taken out of commission. In European waters, although twenty of these "Q" boats were sunk in action, they had in turn sunk thirteen submarines. The North American schooners were less successful as marauders, but more fortunate in that they survived.

Lunenburg in the late 1920s.

Alas, when the war at last mercifully ended, the boom times were over for the Lunenburg banks fishing fleet. Fewer and fewer vessels sailed out of the harbour over the years, and by 1925 the number fell below a hundred, never to rise higher. By 1929, wrote Balcom, the number had fallen to 71, and to 26 by 1932. As Mike Parker put it in *Historic Lunenburg*, "For all intents and purposes, the classic salt banker had by this time sailed into history."

The fleet had prospered during the war years, with big catches and high prices, but at the end of the war prices dropped sharply, though not the cost of vessels and outfitting. An item in the *Bridgewater Bulletin* on January 3, 1928, stated that the cost of a totally equipped fishing schooner almost doubled between 1913 and 1926. The industry was affected too by the worldwide depression that began in 1921, with prices for Lunenburg dried cod declining from $12.65 per quintal in 1919 to $6.40 a quintal in 1921, according to Ruth Fulton Grant. After that date, she claimed, the industry ceased to operate on a profitable basis. There was also stiff competition after the war from Norway, Iceland, and Britain. The Norwegians especially, nearer to their supply of fish, were producing better quality fish for better prices.

Both Nova Scotia and Newfoundland lost markets in the south to the Europeans. The Lunenburg fleet was protected to some degree because their heavily salted "cure" remained the preferred one in parts of the West Indies; indeed, in some places, said Grant, it brought higher prices than the superior Newfoundland, Gaspé, and Norwegian products. However, that advantage disappeared when Newfoundland began producing the Labrador "slop cure" comparable to the Lunenburg product. By 1924 Newfoundland was providing about half of Puerto Rico's requirements for salt cod, practically all of which had previously been supplied by Lunenburg.

The *Lunenburg Argus* of October 16, 1924, was pessimistic, noting that the fleet had decreased to an alarming extent, and even then the catch was small for the number of vessels engaged. The *Bridgewater Bulletin* on December 22, 1925, observed that the Lunenburg fleet had declined from 110 sail in 1915 to 76 in 1925. Conditions continued to deteriorate throughout the 1920s, and, as Balcom concluded, "nothing the Lunenburgers attempted in the way of remedial action appeared capable of generating growth or even stability within the salt fish industry." In 1921 the "frozen bait" trip was introduced, thus lengthening the fishing season. Five vessels left in early March before the regular spring voyage and returned with a good catch, and next year many more vessels made the trip with much success. Despite such measures as importing crews from Newfoundland the general decline continued. The vessels as well as the men in the 1920s were finding alternative, more lucrative employment; the *Maritime Merchant* in May 1924 claimed that there were seventy-five schooners engaged as rum-runners, as many vessels as in the fishing fleet itself.

The "foreign Protestant" farmers of 1753 had become fishermen. In a reversal of roles, by the twentieth century it was agriculture that had taken a back seat to the fishing industry. "Lunenburg has more money in the Savings Bank, per head of population, than any other town in Eastern Canada," declared the *Canadian Fisherman* in August 1915. "The soil in the neighbourhood is fertile but farming is not extensively carried on. Everybody is in the fish business."

The Lunenburg banks fishing fleet had been through glory days and through tough times, through prosperous wartime years, then decline. But it had survived, and would survive, in one form or another, for many years to come. In the mid-1920s a "forest of spars" still graced the harbour, until the schooners once again headed out to the banks.

an ISLAND in a STORMY SEA

> Tomorrow's tide is deaf to call,
> Without recourse the daylight dies;
> No treaty binds the shifting squall,
> With wind there is no compromise.
> —Charles Bruce, "Tomorrow's Tide"

THE NORTH ATLANTIC OCEAN IS A STORMY PLACE. Tropical depressions, tropical storms, hurricanes. And gales. All have visited Atlantic Canada, and all have been deadly.

These storms generally form in the tropics because of the hot, humid conditions over the waters there. In the northern hemisphere winds in these storms circulate counter-clockwise near the earth's surface. By the time they reach the colder waters the storms have often become extratropical, their cyclonic motion generally having straightened out, but they can remain very powerful. At sea and on land they can be devastating. As it says in Job (37:9), cold comes from the north, but the whirlwind comes out of the south.

Hurricanes are one kind of a tropical cyclone, the fiercest. Because of their great size and the strength of their winds, they are the costliest and most destructive. They combine strong winds, tornadoes, heavy rains and storm surge leading to flooding, resulting in extensive damage and many deaths. A hurricane is a tropical cyclone with maximum sustained winds of at least 119 kilometres an hour, or number 12 on the Beaufort Wind Force Scale. One theory is that the word "hurricane" came, via the Spanish word *huracán*, from the name of the Carib Indian god of evil, which in turn stemmed from the Mayan weather god Hurikan, one of their creator gods. Thus "hurricane" essentially means "evil wind."

Although not as devastating as full-blown hurricanes, the lesser storms can also be dangerous. A tropical storm is a cyclone with maximum sustained surface winds of 63 to 118 kilometres an hour, whereas a tropical depression, with winds of 37 to 62 kilometres per hour, typically does not have the organization or the spiral shape of the more powerful storms. A gale is defined in dictionaries as "a very strong wind." Gale-force winds, numbers 8 to 10 on the Beaufort scale, blow from 62 to 102 kilometres an hour. Sea conditions under number 10, "whole gale," are described as: "Very high waves with overhanging crests. Large patches of foam from wave crests give the sea a white appearance. Considerable tumbling of waves with heavy impact. Large amounts of airborne spray reduce visibility." Not a hurricane, but nevertheless not a storm you'd want to experience on the open ocean.

The North Atlantic hurricane season runs officially from June 1 to November 30, though Environment Canada maintains that about 85 per cent of the tropical cyclones that actually make landfall occur between August and October. Memorable storms that struck Atlantic Canada in the month of August in the past made such an impression that in parts of the region all late summer storms became known as "August gales."

AN ISLAND IN A STORMY SEA

Some fifty years before the calamitous gales of the 1920s, a severe storm of hurricane strength swept over Cape Breton Island and the eastern mainland of Nova Scotia on August 24–25, 1873. Known as the Great Nova Scotia Cyclone, it seems to have been the first North Atlantic storm actually to be referred to as "the August gale." It was a typical Cape Verde–type hurricane, spawned by tropical waves that were formed in Africa during the wet season, eventually moving out over the sea, increasing in strength as it moved westward over the warm tropical water towards the Caribbean. As often happens, on approaching the western side of the ocean it curved northward, eventually ending up in the Maritimes and Newfoundland. Newspaper correspondents of the day frequently labelled it "a perfect hurricane." Tragically, meteorologists in Toronto knew a day in advance that the storm would make landfall in the Maritimes, but no alarm was raised because the telegraph lines to Halifax were down.

It's ironic, if not somehow poetically fitting, that the winds that begin their life in faraway Africa should end up in Atlantic Canada, for before the supercontinent of Pangaea broke apart and North America drifted away from Africa some 250 million years ago, thus creating the Atlantic Ocean, Nova Scotia nestled up close to Morocco, and the Avalon Peninsula in eastern Newfoundland was most likely part of the African continent. A bizarre kind of reunion. It seems appropriate too, perhaps, that as the winds approach Atlantic Canada in such storms they replicate their more benign African beginnings. As they move north into the higher latitudes of the Atlantic they weaken fairly rapidly, and their structure changes: they increase in size, having a larger wind field, particularly on the right side of the storm track, and they're weaker at the centre. But they can be much bigger storms, impacting larger expanses of sea and land. They also pick up speed, thus moving faster through the area.[1]

Sometimes, the storms do not weaken enough. When rain began to fall that late August day in 1873, most Cape Bretoners expected nothing more than an ordinary late summer storm. By nightfall, however, the great gale was raging, and by the next afternoon destruction across the island was widespread. Hundreds of homes had been destroyed, churches damaged, bridges washed away. Barns, outhouses, and trees blown down. Wharves, dykes, and fish stores ruined. Vessels in port were dragged ashore or demolished by running into docks or other vessels. Indeed, the wreckage of ships littered the coast. At Port Morien, then known as Cow Bay, the harbour was full of ships engaged in the coal trade; the breakwater and loading facilities were badly damaged, many of the vessels driven on shore. At Petit-de-Grat, on Isle Madame in the south of Cape Breton, the schooner *Ocean Wave* was riding at anchor when the storm set in; at the height of the gale she was dragged out of the harbour and smashed to pieces on Cape Hogan. All on board perished. Across the strait on the mainland, in Canso, about a hundred buildings were damaged, a new church was blown into a lake, a schoolhouse and a cannery demolished.

"We regret to have to chronicle the heaviest gale within our recollection on our coast," declared the *Cape Breton Times* of Sydney on August 30, 1873. "The greatest gale experienced since 1810 swept over the Island of Cape Breton on Sunday last," concurred the *North Sydney Herald* on August 27. "Our usual November winds and waves have not equalled, this century at least, the terrific, magnificent yet awful blast.... The harbour, which we had fondly imagined so landlocked that no winds or sea could disturb us, contains to-day the wrecks of 25 ships.... Travelling is stopped on account of broken-down bridges.... The crops throughout the country are damaged most fatally."[2] In his book *In Peril on the Sea* Robert Parsons quoted another *Herald* reporter as saying "A gloomier and wilder night no one can imagine." He went on to describe the scene: "Vessels riding over the best mooring ground in the harbour were rapidly drifting ashore, dragging great anchors weighing 2,000 pounds....The howl of the storm drowned the loud cries of the shipwrecked sailors scrambling up the cliffs or in terror clinging to the rigging. Monday morning presented a gloomy spectacle. A barque's mainmast here, a schooner's foremast there, booms, bowsprits, mainsails, staysails, mixed up with what was once running and standing rigging."

Estimates of the number of people killed in the storm vary widely, from almost a thousand to less than half that number, but clearly too many people died. And most died at sea. Several sources agree that about twelve hundred ships were sunk or smashed. Vessels arriving in port reported large quantities of "wreck stuff" at sea.

Although the storm caused the most severe damage in Cape Breton, it was not the only part of the Maritimes to be affected. Prince Edward Island, which had become a province of the Dominion of Canada only the previous month, also suffered considerable damage. As Edward MacDonald wrote in "The August Gale and the Arc of Memory," Islanders were somewhat casual about tropical storms, believing that the cold water of the North Atlantic and "the massive breakwater of Nova Scotia" would protect them from hurricanes. But this time it was different. Buildings, wharves, and vessels along both sides of the Northumberland Strait were destroyed by an unprecedented storm surge. MacDonald described the 1873 hurricane as "the second deadliest storm in Island history," after the Yankee Gale of 1851, noting that vessels caught at sea in the gulf were irresistibly swept south "towards the August Gale's deadliest coast, the North Shore of Prince Edward Island."

Eric Allaby, a New Brunswick politician from Grand Manan and marine historian, prepared, a century later, a long list of the shipping losses in the August gale of 1873. He maintained that this gale was "[p]ossibly the most destructive storm within written history, in the general area of the Atlantic Provinces," for it affected the shipping of all of the provinces as well as New England. The greatest damage was in Cape Breton and in eastern Nova Scotia, but ships were lost or wrecked as well off Prince Edward Island and the Magdalen Islands in the Gulf of St. Lawrence.

On September 14, 1873, some inhabitants of Little Codroy River, on Newfoundland's west coast, discovered the hull of a vessel floating about five kilometres offshore. They boarded the wreck, which turned out to be the *Three Brothers* from Petite Riviere in Lunenburg County, registered at Lunenburg. The boat was loaded with fish. Having towed her ashore and pumped her dry, they found the bodies of five men in the forecastle in an advanced state of decomposition. Dressed in their oil clothes, they had evidently prepared for heavy weather, and had doubtless met their fate in the terrible storm of August 25. A small sum of money found on board, $5.25, was used to bury the dead. Other articles found in trunks were twenty-two shillings, a handkerchief and stockings, a Bible, and a book of Protestant hymns, almost destroyed by the water. A boy's boots and clothes were found, but no body.

A similar story was told of a small schooner, the *Good Intent*, from Arichat, on Isle Madame. The vessel was towed into the harbour at Port Hood on the west coast of Cape Breton, brought to the wharf, and righted. Inside the cabin were the bodies of the crew of five men and two boys. "It would appear," the *Halifax Morning Chronicle* reported on September 4, "that the crew, when they had given up all hope of saving themselves, repaired to the cabin, to engage, very probably, in devotional exercises, and there met their sad end. They were Frenchmen."

Several small ports on Nova Scotia's South Shore also suffered losses. The *Young Nova Scotian* of Vogler's Cove, with a crew of thirteen sailing from Labrador to Halifax, was last seen on August 21 off the Bay of Islands on the west coast of Newfoundland. The *Thetis*, a newly built LaHave fishing schooner owned by Corkum

& McKean of Bridgewater, was found loaded with codfish and boarded at New London, Prince Edward Island, just after the storm, dismasted and waterlogged. Pumped out and taken into the harbour, the bodies of two men and a boy were found aboard. Twelve of the ill-fated crew came from two families, the Corkums and the Shankles.

According to the *New York Times* of September 28, 1873, four vessels from Nova Scotia's Queens County, along with their crews, were lost in the "memorable" August storm. Among them was the schooner *Leader* of Port Mouton, which had sailed from Liverpool for Sydney, and was lost off Sambro on the fatal Sunday. The master, Isaac Smith, left a wife and a small child; the three crewmen were all young men and unmarried. The *Liverpool Advertiser* reported the loss with all hands of the schooner *Bonnie Jean* on the north side of Prince Edward Island. She had sailed from Port Medway three weeks previously on a fishing voyage. The crewmen were "all young men, the only support of widowed mothers and aged parents, who have thus been plunged into the greatest distress imaginable."

Many ships and men from the New England fleet out of Gloucester were also lost. The schooner *Angie S. Friend* was wrecked and washed ashore at Port Hood, with 14 men lost. The *Henry Clay* left the Grand Banks three days before the storm and foundered with all hands on the passage home, 10 more men gone. During the gale there were apparently more than a hundred American ships fishing in the Gulf of St. Lawrence, especially around the Magdalen Islands. A correspondent from Gloucester in the *Halifax Evening Reporter* declared that the gale, because of the disabling of the fleet and insurance costs, was a much more serious calamity to businesses in the town than the conflagration of the previous week that had consumed five blocks of stores in the business centre. George Procter, who published *The Fishermen's Memorial and Record Book* that year— despite the fact that some of the pages in type and even part of the manuscript had been lost in the fire—estimated that nine vessels from Gloucester had gone down with all hands. Along with men swept overboard from other vessels, the number of Gloucester men lost was 128, many of the "very best skippers and smartest fishermen" of the port.

The August gale of 1873 wrecked more than ships and buildings. Many trees were uprooted, whole forests flattened. The hopes of farmers in Guysborough County, on the eastern Nova Scotia mainland, were ruthlessly destroyed, said the *Halifax Evening Reporter* of August 29, for heavily wooded hills that had survived all previous gales within memory now "resembled newly mowed fields of hay.... Everything of a movable nature has been twisted and turned and tumbled about as if an army of guerillas had ransacked the country." Crops were damaged, and the blasting of the wind scorched leaves like frost, turning them brown.

Dave Thurlow, on CBC Radio, Charlottetown, in 1998 noted that late summer storms on Prince Edward Island can offer "an unusual bounty": hundreds, if not thousands, of live lobsters thrown up onto the shore by the force of a storm. "In 1873," he said, "when lobsters were really plentiful, the numbers that ended up on the shoreline were staggering. Old records say there were 1,000 lobsters for every 11 yards of coastline—and they were piled up to five feet deep."[3] Perhaps hyperbole here?

The correspondent of the *Saint John Tribune* reported a minor incident of the storm with some glee, providing a bit of very light humour in the midst of all this grimness. While the storm was at its worst at Pointe-du-Chêne, near Shediac, New Brunswick, about 1:00 or 2:00 A.M., with the sea even washing away the foundations of some of the houses, "two young ladies (of that class who have a great taste for the awfully grand, and who even look on the scene presented with admiring eyes)—so the story goes—got up to watch the big waves; and while so engaged the sea was busy washing the inside out of their pantry...and all the preserves they had so toiled over and boiled over to get ready for winter were upset, and the butter and the lard, and all the other little etceteras of a well-stored pantry likewise. Their ideas of the grandness of big storms," the reporter concluded smugly, "have undergone a change."

Eventually, as all things must, "the terrific, magnificent yet awful blast" of August 1873 had come to an end. "Ever since," wrote John P. Parker in his book about Cape Breton ships and men, "a late summer storm over this island has been known as 'The August Gale.'"

"Months That Sailors Dread," declared the headline of an item in the *New York Times* of August 27, 1894. Because tempests arrived so unexpectedly in August and September, Atlantic mariners feared the gales of winter less than "the furious outflies of summer's end and fall's beginning." It was the time of year when the weather "runs to excesses, changing with great suddenness from the very good to the very bad." The most disastrous August hurricane, the reporter wrote, was the storm of 1873, when eleven hundred vessels had been wrecked.

But August gales have wreaked havoc since sailors first ventured onto North Atlantic waters. One of the earliest August storms we know about is the Great Colonial Hurricane, which swept over New England on August 25–26, 1635, terrifying the Pilgrim settlers who had landed at Plymouth Rock on the *Mayflower* just fifteen years earlier.

The details of one of the earliest recorded August storms in Canadian waters, in 1711, are vague, but clearly something momentous occurred. In 1710 the French colony of Acadia had fallen into British hands with the conquest of Port Royal, but the French were still firmly ensconced at Quebec City. A large British fleet, consisting of nine warships, two bomb vessels, and sixty transports and tenders, under the command of Rear-Admiral Sir Hovenden Walker, left Boston at the end of July and sailed into the Gulf of St. Lawrence intending to attack the French stronghold upriver. The invasion came to a halt when a "violent storm" overtook the fleet on August 22, sending several of the ships crashing onto the rocks at Île aux Oeufs, west of Sept-Îles. Estimates of the number of deaths vary from about 890 to 2,000, including soldiers, seamen, and 35 women who were attached to the regiments. Some sources suggest, however, that while wind was most likely involved in the disaster, the main villain was dense fog. In any case, much of the fleet ended up on the rocks, the invasion failed, and New France was allowed to flourish for another half century. Thus can weather change the course of history.[4]

Later in the century, in early September 1775, a particularly deadly gale that came to be known as the Great Newfoundland Hurricane, devastated the eastern part of the island and the Grand Banks. In the United States the storm is called the Independence Hurricane because it first made landfall on the coast of North Carolina on August 29, 1775, at the beginning of the American Revolutionary War. In fact, it's not certain that this is the same storm that struck farther north several days later. There can be no doubt, however, that a major storm ravaged the waters around Newfoundland that year. The beaches were said to be littered with the bodies of drowned sailors, and for many years afterwards bones washed ashore.

Marine geologist Alan Ruffman, who has studied the Newfoundland hurricane more closely than anyone else, thinks this storm is possibly the most tragic natural disaster in Canadian history, and in the history of the French islands of Saint-Pierre and Miquelon as well. He quotes a report in the *Annual Register* for 1775 that, he says, casts "a particularly gruesome light on the storm":
"[September] 11th. At St. John's, and other places, in Newfoundland, there arose a tempest of a most particular kind—the sea rose on

a sudden 30 feet; above seven hundred boats, with all the people belonging thereto, were lost, as also eleven ships with most of their crews. Even on shore they severely felt its effect, by the destruction of numbers of people; and for some days after, in drawing the nets ashore, they often found twenty or thirty dead bodies in them; a most shocking spectacle!" Ruffman thinks that the effect felt "even on shore" likely refers not to a great loss of people on land but to the loss of men caught by the storm at sea in small boats.[5]

The Great Newfoundland Hurricane may also have an etymological legacy. Many people on shore claimed they could hear the cries or "hollies" (hollering) of the drowning men at sea. The word "holly" in the *Dictionary of Newfoundland English* is defined as the "cries of dead fishermen heard on stormy nights."

Ten years after the terrible August gale of 1873, another early fall storm caused much destruction on the Grand Banks. The *New York Times*, in dispatches from St. Pierre, reported on September 5, 1883, that a cyclone had struck the banks on August 26, followed by another storm a few days later. Some fifty vessels of the French banking fleet arrived in St. Pierre with cable chains, anchors, batteaus, dories, and lines all swept away. There was a great loss of life. Over thirty vessels were missing. There had been no warning; the storm broke furiously on the fleet when most of the dories were away from the ships overhauling or setting trawls. A Gloucester schooner had arrived back in port towing a dismasted and abandoned Newfoundland banker, the *Mediane*; eight of the crew had been saved by a Lunenburg schooner and taken to Halifax. The report said that the storms were "the most violent ever known on the Banks." Six days later, in an item called "Mishaps to Mariners," the *Times* reported that the Lunenburg schooner *Restless* was towed into Barrington full of water and dismasted, with the dead bodies of the crew in the cabin. The schooner *Madonne*, of Catalina, Newfoundland, broke from her moorings on the Grand Banks on

August 30 and drifted across the bows of a French barque, carrying away her mainmast. The captain and three men left in a dory hoping to board the barque and were never seen again. The remaining eight men were taken off by a schooner and landed at Lunenburg. The *Madonne* was later taken into St. John's by the American schooner *W. E. McDonald*. These incidents reveal the international character of the banks fishery, as was the case four hundred years earlier, in the sixteenth century, when Basque, French, Portuguese, and English vessels had coexisted on the banks.

The *Restless*, according to the *Lunenburg Progress* of September 12, 1883, was probably bound for home from Quero Bank when she was caught in the gale. The vessel had been built at Vogler's Cove in 1881 and was owned by the Reinhardt brothers of LaHave. Years later, the members of the LaHave Life Guard Lodge, the Independent Order of Good Templars, held in their hall at Getson's Cove "an entertainment consisting of vocal and instrumental music, dialogues, recitations, etc., for the purpose of raising funds to erect a monument in memory of members of the lodge lost at sea in the great gale of August 1883." Over two hundred people had attended, the *Progress* reported on October 24, 1888. The event likely honoured the men on the *Restless*, since the vessel was owned in LaHave and Captain Daniel Getson was a member of this temperance lodge.

Another Lunenburg schooner, the *Welcome*, ran into serious trouble on August 29, 1883, in a gale off the northern coast of Prince Edward Island. When the seas began to run high, the crew of ten men tried to round East Point to return to the safety of the harbour at Souris, but they were struck by a huge wave, which hurled the vessel onto her beam ends. Israel Spindler of Lower LaHave, the only survivor of the disaster, saved himself by clinging to the wreckage for twenty-six hours. He had been washed overboard, and when he surfaced he heard the men in the cabin, including his brother Gabriel, cry out once, "and then their voices were all hushed in death." Three others were clinging with him to the rigging, but one by one the exhausted men were washed away by the seas. The managing owner of the *Welcome* was Benjamin Himmelman of Middle South (now Bayport). Three of his sons—Albert, who was the skipper, Eli, and young George—drowned that day, along with their two cousins Henry and Stannage Himmelman. Spindler had tried several times to help George, but the "last time, he looked up into my face and smiled, then sank to rise no more." Later, the ship was salvaged by Islanders, and they found the bodies of the five crewmen trapped in the cabin and forecastle, "all young, the oldest only reaching the prime of life." During the same gale, Benjamin Himmelman was aboard the steamer *Edgar Stuart*, which rescued the schooner *Active* off Betty's Island near Prospect and towed her into Halifax. The *Welcome* eventually went back to sea, after a lengthy battle over ownership and the payment of salvage costs.[6]

August gales continued to unleash their fury over the years. In 1892 the LaHave schooner *Cashier* foundered on the banks in the tail end of a tropical hurricane while on an August fishing voyage. James Forsey of Grand Bank, Newfoundland, master of the fishing vessel *Mary F. Harris*, survived the storm but witnessed the loss of two nearby vessels. He knew the storm was coming but his only recourse was to ride it out at anchor. At daybreak he found the *Cashier*, still lying at anchor but partially submerged, waterlogged and helpless. Another schooner, the *George Foote*, from Forsey's hometown, with seventeen men on board, many of whom he knew well, had disappeared. He discovered buoys, trawl tubs, broken dories, and other floating debris marked and identified as belonging to the *George Foote*. The two vessels must have dragged anchor during the storm and collided, Forsey concluded. The *Cashier* had nineteen men on board, from the LaHave area, West Dublin, and Vogler's Cove in Lunenburg County. "Once again," declared marine historian Robert Parsons, "a Nova Scotia town that sent a fleet of ships down to the sea had paid a heavy price for the riches of the ocean."[7]

A year later, on August 21–22, 1893, another large storm hit Nova Scotia, referred to by some as the "second August gale." The *Halifax Morning Chronicle* reported on the "roar of the velocity of the wind, the falling wires, the tumbling fences and signboards, the breaking of windows, the snapping of branches and crashing of trees, the ringing of bells and the shaking of buildings—it was uproar gone mad." Damage occurred all over the province. The next day houses were blown down in St. John's, Newfoundland, and there were reports of heavy loss of life out on the banks.

In the midst of this violent, stormy sea lies a sizeable, rather placid-looking sandbar called Sable Island. There are sand beaches, sand dunes, sandspits, and sandbars. "Sable is so totally a place of sand," remarked Bruce Armstrong in *Sable Island*, "that any other name for this island would seem unthinkable." But there's plenty of life on Sable. Although the winters are bleak, grey and windswept, in other seasons there are beach peas and sandwort, goldenrod, iris, daisies, yarrow, wild roses and juniper, pearly everlasting, bayberry and crowberries, even tiny orchids and blue-eyed grass. Wild blueberries and wild strawberries tempt the palate. Ripe cranberries sparkle like drops of blood amid the greys, pale golds, and soft greens. The island's climate is generally milder than the mainland, according to Zoe Lucas, who spent many years there. There are seals, both grey seals and harbour seals, flopping about on the beaches, and ducks in freshwater ponds bordered by bulrushes and wildflowers. In the sky above are innumerable gulls and terns, and many other species of birds. And then there are the horses, feral horses, with their long manes and equally long tails flying in the wind, roaming the island in small herds and feeding on the thick marram grass on the dunes.

How could such an idyllic-sounding paradise become known as "the graveyard of the Atlantic," "a sinister scimitar of sand," an "island of sand and ruin"? Wrote Joseph Howe in his poem "Sable Island":

AN ISLAND IN A STORMY SEA

> Dark Isle of Mourning—aptly art thou named,
> For thou hast been the cause of many a tear;
> For deeds of treacherous strife too justly famed,
> The Atlantic's charnel—desolate and drear...

The trouble with Sable, of course, was that it was there, lurking ominously in the middle of the wide ocean. Ships tended to run into it.

Sable Island is located about 160 kilometres from the nearest landfall in Nova Scotia, near Canso; it is approximately 300 kilometres southeast of Halifax, 187 nautical miles from Lunenburg. Shaped like a crescent moon lying on its back, it's more or less 40 kilometres long and about 1.5 kilometres wide. At the east and west ends of the island there are the spits of sand, which change in size and shape, as does the island itself, when modified by the winds, currents, and tides. A "dune adrift," as Marq de Villiers and Sheila Hirtle so aptly described it. Beyond the spits are the treacherous east and west bars, submerged extensions of the island extending like tentacles far out into the ocean.

There are many reasons why Sable Island became known very early on as the "place well-known for shipwrecks." For one thing, because of its flatness and muted colours the island was very hard to see, hidden also by waves, storms, and fog. The bars, just below the surface of the water, were more dangerous to mariners than the island itself. As well, the island was frequently shrouded in dense banks of fog, caused by the meeting of cold water and air from the north in the Labrador Current mingling with the warm Gulf Stream water from the south. The dangers were compounded by the fact that Sable Island lay in the path of one of the major shipping routes between Europe and North America. Hundreds of vessels sailed past each year.[8]

And the fishermen were there because it was a very good place to catch fish. They knew the dangers of working close to the island, but when the weather was fine it was well worth the risk, for the fish

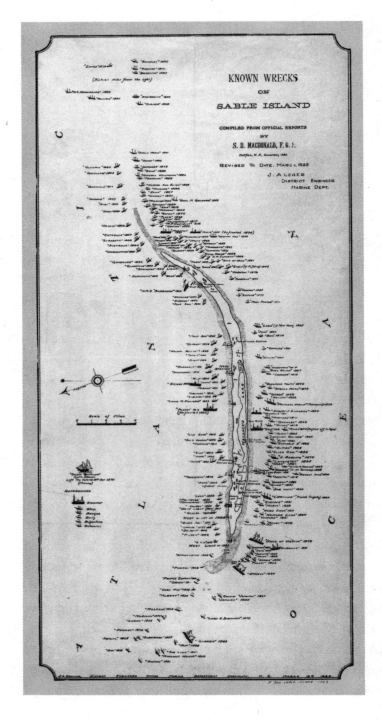

were plentiful. With its sandspits, however, and with its sandbars just beneath the surface of the sea, Sable was a large trap, waiting to catch vulnerable vessels like moths in a spider's web. Captain Angus Walters, according to G. J. Gillespie in *Bluenose Skipper*, opined that "[a]s any sailor knows, Sable Island or anywhere near isn't the place to be when you have a real breeze of wind."

One of the first recorded ships to founder in Sable Island's shoals was the *Delight*, in 1583. Elizabethan explorer Sir Humphrey Gilbert, with three ships, had sailed to Newfoundland to claim the island for the British. Thereafter, he planned to stop at Sable Island because he had heard there might be cattle or pigs available for feed. The ships were apparently caught too close to the island when strong winds reached gale force. About eighty-five men died when the *Delight* went down, though a few managed to get to Newfoundland in a small boat, saved partly by drinking their own urine. Later, the leader of the expedition himself drowned when the *Squirrel* went down north of the Azores. Sir Humphrey was last seen sitting aft with a book in his hand, repeatedly crying out: "We are as neare to Heaven by sea as by land." A few lines of poetry collected by Judge DesBrisay in his history of Lunenburg County echo the sentiment:

'Tis well to find our last repose
 'Neath the churchyard's sacred sod;
But those who sleep in the desert or deep,
 Are watched by the self-same God.

Only the *Golden Hind* made it back to England from this fateful trip. It's notable that the *Delight* was lost at Sable Island in an August gale, some 340 years before the vessels of the Lunenburg fleet foundered in the 1920s.

The *Delight* was just one of hundreds of ships that met their fate on Sable Island. As Harry W. Hewitt observed in his history of the town of Lunenburg, written about the time of the First World

War, the "bones of many a Lunenburg fisherman lie embedded in the shifting sands of Sable Island, as do the timbers of many a Lunenburg vessel." A great many wrecks, of course, were never recorded. But on shipwreck maps of Sable Island the ships are stacked so close together all along the shores that one imagines wreckage would be piled sky high along the beaches to this day. The elements soon take care of the debris.

In 1801 the Sable Island Humane Establishment was set up to help rescue survivors. For the first time there were permanent residents on the island. The Canadian government assumed responsibility for the life-saving station at the time of Confederation in 1867, and soon thereafter two lighthouses were constructed, one at each end of the island. By the 1890s telephones allowed the life-saving crews and lighthouse keepers to co-ordinate search and rescue operations. Government steamers called regularly at the island, bringing supplies. By 1895 there were five life-saving stations stretched out along the island. While much was done, therefore, to help the victims of wrecks, beyond the building of lighthouses little progress was made in preventing shipwrecks in the first place. During the time of the Humane Establishment, from 1801 to 1958, "some 222 recorded schooners, barques, brigs, brigantines, ships, terns, pinks, and steamers came to grief."

In 1905, according to Bruce Armstrong, telegraph communication was established between Sable Island and the mainland, thus finally breaking down the island's isolation. Added to the island's small population was a new breed of man, the wireless operator, or "brass pounder," who communicated with passing ships and with stations on the mainland. The Nova Scotia writer Thomas H. Raddall, who was sent to Sable Island in 1921 by the Canadian Marconi Wireless Telegraph Company, felt that he was facing a year's exile on an island "full of wrecked ships and dead men's bones, a desert in the sea." At that time there were about forty people living on the island, the crews of the life-saving stations, the two lighthouses, and the wireless station—twenty men, some of whom had wives and children. "I learned for myself," Raddall revealed in his memoirs, *In My Time*, "the strange fascination of this boneyard in the sea, where a long gale from an unusual quarter would expose old wreckage buried in the dunes, or the remains of a rude hut made from ship timber by castaways, or human skulls and bones that had been covered perhaps for centuries." His opinion of the island, where, he complained, nothing grew higher than his knees, did not change much. On his last night on Sable he wrote a piece of doggerel that ended: "And when the Devil lets me into Tophet with a curse, / I'll tell him, "Nick, it ain't so bad, I've seen a place that's worse!"

Sable Island may have had telegraph communication with the mainland, with steamers, and with the sleek transatlantic liners that skimmed over the ocean waves, but there was no way to contact the fishing vessels. They had no wireless, no reliable weather forecasts, most had no engines. Captains had to rely on a barometer, often called "the glass," and their own judgment to warn of approaching storms. The fishing schooners were at the mercy of the stormy North Atlantic.

The twentieth century got off to a bad start when a major hurricane laid waste to faraway Galveston, Texas, in September 1900, killing at least 8,000 people. The storm moved north to the Great Lakes, bringing strong winds to Toronto and destroying crops on the Niagara Peninsula. Heading eastward, it lost strength over Quebec but regained momentum over the Gaspé Peninsula and northern New Brunswick as it headed out over Atlantic waters towards Newfoundland. On September 12 the centre of the storm was off Port aux Basques. Most of the damage and loss of life occurred at sea. On October 4 the *Western Star* of Corner Brook confirmed the loss of eighty-two schooners ashore or foundered plus another hundred seriously damaged. The fishing fleet of Saint-Pierre and Miquelon was decimated: nine schooners were lost with 120 men, leaving 50 children fatherless. One of the capsized schooners, the *Ali Baba*, was towed into port at St. Pierre with thirteen bodies inside so bloated that only one was recognizable.[9]

As a prelude to the storms that would come a couple of years later, an August gale struck Atlantic Canada in 1924. The storm swept over Nova Scotia from Cape Sable to Cape North on the night of August 26–27. Gardens and crops were battered. In Lunenburg many large trees were uprooted and shipping in the harbour suffered damage. The tern schooner *Richard B. Silver* broke away from her moorings, ran afoul of another vessel, then smashed part of the wharf and store at the outfitting company Adams & Knickle.

On September 4, 1924, the *Lunenburg Argus* maintained that the severe gale of the previous week had been the worst in years. The *Marie A. Spindler*, Captain Willett Spindler, had been fishing on the banks some distance northwest of Sable Island. In the storm she lost her cable and two dories, her main gaff was broken, rails were smashed and the windlass torn out. Crew member Gordon Tanner of Stonehurst was swept overboard at the height of the storm. The *Jean Smith*, Captain Albert Selig, was also damaged and compelled to head for her home port. On this vessel Maurice Hiltz of Martin's River had gone overboard to his death. Both vessels arrived in Lunenburg with their flags at half-mast.

The *Halifax Herald* on September 5 reported that the schooner *Elsie M. Hart*, Captain Frank Meisner, had arrived in Lunenburg from the banks a few days earlier, having lost two of her crew. It was the third fishing vessel in four days to sail into the harbour past Battery Point with its flag lowered.

A vessel's flag at half-mast was a poignant sight, all too familiar to the people of North Atlantic fishing ports. "'How often, oh, how often' has the signal referred to in the following lines of poetry been seen," wondered Judge DesBrisay in his 1890s history of Lunenburg County:

> Half-mast high the signal floats!
>   She's coming in from sea;
> Some sailor of her crew is gone,—
>   Who may the lost one be?
> The landsmen gaze as she draws nigh,
>   With trembling sad concern,
> The vessel's name to learn,
>   That comes with colors half-mast high.

Each year the fleet had sailed for the banks, and over the years many men had been lost. On the black granite columns of Lunenburg's fishermen's memorial, shaped like a compass rose and located on the waterfront, are the names of more than three hundred fishermen who perished at sea before 1925. On a Sunday afternoon towards the end of September 1925, a unique memorial service was held on the wharves of the outfitter Zwicker & Company for the men who had been lost at sea that year. Four of the men were from Lunenburg: Richard Hynick died on the schooner *Douglas J. Mosher*, Freeman Feener on the *Jean Smith*, Robert and Richard Schnare on the *Mary H. Hirtle*. Two Newfoundlanders, William Newport and Silas Grandy, were lost from the schooner *Vera P. Thornhill*. The idea of a service to honour the dead fishermen had come from Rev. W. Ryder of St. John's Anglican Church, who had approached the town council proposing that a day be set aside each year. There had been memorial services in churches for members who had been lost, but never a public service. Thousands poured into town to pay tribute to the six men who had died. Flags were flown at half-mast from the vessels in port and from the outfitting companies. The band led a procession from the courthouse to the waterfront. Wreaths were cast on the waters of the harbour. "It was a beautiful and most impressive tribute," said the *Lunenburg Progress-Enterprise* on September 30, "the sea bearing on its bosom the wreaths and flowers of remembrance, which drifted with the tide rising and falling in unison with the singing by the choirs massed on the harbor front." In its November 1925 issue, the *Canadian Fishermen* observed it was strange, in this port of fishermen, that this was the first service of its kind ever held. It was but a dress rehearsal for memorial services in the years to come.

Despite the losses over the years, however—and through almost 175 years of history—it has to be said that the town of Lunenburg, tucked cozily beside its sheltered harbour at the head of Lunenburg Bay, had come through relatively unscathed. Lives had been lost, to be sure, too many lives, but there had been no major disasters. Lunenburg's time to mourn in earnest would come when the August gales confronted the island of sand.

the AUGUST GALE of 1926

Lunenburg schooner *Sylvia Mosher* total wreck on outer bar, north side, near Number Four station, lying on side. No sign of crew.

THESE CHILLING WORDS, sent in a radiogram from Superintendent H. F. Henry of the Sable Island life-saving station on the morning of August 10, 1926, to C. H. Harvey, local agent of the Department of Marine and Fisheries in Halifax, were the first indication that the recent storm had been deadly. After the discovery of the wreck, Henry later reported, the life-saving staff had spent all day searching the beach for the crew, "scanning the billows that rolled in over the sandy shoals." Many dories had come ashore on the island, most damaged beyond repair, but the men were nowhere to be seen.

A stranger in the busy town of Lunenburg, declared the *Bridgewater Bulletin* of March 16, 1926, would surely have wondered what holiday was being celebrated, for flags were waving from all the buildings of the outfitting companies and from the topmasts of the schooners in the harbour. The occasion was in fact the departure of the Lunenburg fishing fleet, on March 10, off on its first trip of the year to the banks. Seldom had so many vessels, about forty, left the port on the same day, and a pretty sight they made, declared the reporter, as they sailed out into the bay beyond Battery Point, their white sails gleaming in the winter sunshine. Carloads of bait in the form of fresh herring and squid had arrived by train and been loaded aboard, along with ice, salt, and provisions for the crew; but departure had been delayed for six or seven days because of stormy weather. Most of the rest of the fleet would leave the next day. The *Canadian Fisherman*, in its April 1926 issue, thought there would be about ninety vessels in the fleet that year; some four hundred fishermen had come from Newfoundland to make up the crews, and still men were scarce for the job. Among the vessels was a new schooner, the *Mahala*, whose captain, Warren Knickle, only twenty-five years old, was the youngest captain in the fleet. The much more famous *Bluenose*, champion in the international fishermen's races, also departing that day, somehow managed to get stuck on a mud shoal on the way out; fortunately, after only about an hour, the pride and joy of Lunenburg was freed from this ignominious position.

Perhaps it was an omen, for the first trip of the year didn't go quite as smoothly as was hoped. According to an account in the *Lunenburg Argus* of April 8, two successive storms swept over the fishing grounds in late March, causing the vessels of the fleet to limp back to port, battered and bruised, with loss of gear and anchors, damage to sails and rigging, and in some cases dories had been swept off the decks. The *Bluenose*, fishing off Sable Island, had had a stiff battle with the wind and waves, Captain Angus Walters reported on his return; two of her anchors were gone, the foresail damaged, three hundred fathoms of cable and fishing gear lost. The *Bridgewater Bulletin* of April 6 estimated the damage for the entire fleet at more than fifty thousand dollars. Captain Walters said the *Bluenose* had "spoke" the schooner *Glacier* on the way home, and her captain had shouted across the water that the decks had been swept clean by the storm, everything above deck lost.

On April 29 the *Argus* maintained that *Bluenose*'s "splendid weatherly qualities" had saved her, five days earlier, in yet another storm off Sable Island. She had been caught in the teeth of a "hurricane" that was steadily driving her onto the treacherous Northwest Bar, but she had "won out in the greatest race of her career, a race with Death." A huge sea, described by Captain Walters as a "grandfather sea," had mounted the deck and parted her cable, smashing her starboard boats, carrying away fourteen stanchions as well as part of the rail and bulwarks. The skipper lashed himself to the wheel, and struggled for some six hours to keep the schooner off the bar. Anchored in twenty-seven fathoms of water when the cable parted, the vessel began drifting towards shoal water. Clem Hiltz, a crew member aboard, told Ralph Getson years later at the fisheries museum in Lunenburg that when they sounded the lead and called back eleven fathoms, the captain told them to throw it on the deck and not heave it again; there was sand in it, and it was already apparent they were far too close to shore, with disaster looming. The earlier storm, in March, had been a gruelling experience, one of the worst Lunenburg fishermen had encountered, Captain Walters declared, but this storm was even fiercer. It had snowed all day, "with a breeze of wind from sou' sou' west to really whip her up," and then struck with fury that night. "Don't know as any other vessel could have done it."[1]

In spite of the adverse conditions, however, the *Bulletin* reported on July 6 that the fleet's catch of 65,000 quintals of fish on the spring trip was considered "very satisfactory."

The *Canadian Fisherman* noted in July of 1926 that the Lunenburgers were off for their third trip of the year. The first two trips had been unusually rough and stormy, and much gear had been lost. Seventy-two vessels had sailed on the first trip and seventy-five on the second trip, compared with fifty-three and fifty-nine respectively in 1925. In August the journal pointed out that the handline vessels generally went only on two trips annually and they usually fished in the western banks. Thirteen handliners had gone on the year's third trip. Adding these to the seventy-five vessels in the trawl fleet away on the Grand Banks, there were eighty-eight vessels in the fleet altogether, an increase of fourteen over 1925.

H. F. Henry's cryptic message from Sable Island on August 10 certainly changed the mood of quiet satisfaction. Until then the *Lunenburg Progress-Enterprise* had been much concerned to support, with great enthusiasm, its Liberal candidate in the federal election to be held on September 14; William Duff, Lunenburg fish merchant, would lose in the end to Bridgewater lawyer William G. Ernst, strongly supported by the rival *Lunenburg Argus*. There were other items of interest in the town's newspapers. A new Governor General, Viscount Willingdon of Ratton, had arrived from Britain on the *Empress of Scotland*. Film star Rudolph Valentino, the "Latin lover," had died. And one couldn't help but notice an intriguing item about fashionable women in Paris who had taken to wearing ankle lamps to avoid puddles and to help them step daintily out of motor cars. Locally, there was the ever-fascinating topic of rum-running (sometimes known as "bottle fishing"); though faced with strict temperance legislation at home, Bluenose seamen, with some help from St. Pierre, were none the less doing their noble best to quench the thirst of their desiccated American neighbours.

The weather on land the night of August 7 and the following day had been stormy, but nothing to warrant undue alarm for the men

THE AUGUST GALE OF 1926

at sea.[2] The little village of New Harbour in Guysborough County, near Canso, had suffered the worst when "a storm of hurricane fury" lashed the coast. Some twenty to thirty men there were preparing to go swordfishing, but their boats and gear had been ruined by the tremendous seas that drove into the unprotected harbour. Elsewhere the gale was even seen as a "welcome visitor." Pictou and New Glasgow residents felt the rain would help the crops and the pastures, but even more importantly it would help put out the forest fires that had been raging in the area. The wet weather was also seen as a boon to the crops in the Annapolis Valley. A. E. McMahon of Kentville, president of the United Fruit Companies of Nova Scotia, said the rain that fell was "worth a million dollars" to the apple and potato crops. In Truro the "drenching downpour" helped to fill the reservoir after a prolonged drought; it had been so dry that citizens had been forbidden to use their garden hoses. In Yarmouth, said the *Halifax Herald* of August 9, the storm was "just an ordinary summer's rain."

But on August 11 the *Progress-Enterprise* announced that twenty-five Nova Scotia fishermen on the *Sylvia Mosher* were missing at sea, and more than a hundred relatives were waiting with fearful anxiety for news from Sable Island, "the mariner's dread." The storm in the Atlantic had washed six dories ashore, but the fate of the wrecked schooner's crew was still a mystery. The citizens of Lunenburg County alternated between hope and fear, as scanty reports came in. The vessel had sailed for the banks on July 12 with twenty dories. Had the crew been lost in the great gale of last Saturday, "caught in a welter of wind and sea"? There was some hope that the men had got away in the remaining fourteen dories and were making their way back to the Nova Scotia coast. Or perhaps they had been picked up by another vessel?

It was soon all too apparent that the *Sylvia Mosher* had foundered. "What little flame of hope had flickered in the hearts of men and women in Lunenburg County throughout the day," lamented the *Halifax Morning Chronicle*, "had gone entirely out when night came and there was no word from the great void of waters that held within it the mystery of life and death." Hope gave way to "grim courageous resignation, that spirit of stoic fatalism, so characteristic of men who go out to the sea in ships." Business was at a standstill in the little villages, where families waited for news of the lost crews—"nights of sleepless anguish, days of alternating fear and hope." "The oldest fishermen here," said Gordon Emerson Romkey, warden of the municipality, "men of years of experience of the sea, say there is nothing to do but meet the situation.... What makes the loss so poignant is the realization that the best and brightest of our youth are gone." Being a handliner, the *Sylvia Mosher* took the pick of the young men, "the brightest, ablest, finest men of the community."

Starting out to search for wreckage on Sable Island.

The *Bridgewater Bulletin* on August 17 declared the wreck of the *Sylvia Mosher*, now ashore on Sable Island, to be "one of the worst sea tragedies in the history of the Lunenburg fleet." Each year the sea exacts it toll, but seldom one vessel and all her crew. The *Progress-Enterprise*'s editorial on August 11 spoke of "A Community Loss": the whole county was staggered by one of the fleet "cast away on the treacherous sands of this ocean graveyard." The schooner was only two years old, it was equipped with an engine as well as sails, and it was regarded as the safest of vessels.

The *Halifax Herald*, too, wrote of the *Sylvia Mosher*'s "fatal encounter with the wild Atlantic." "This time the county of Lunenburg is called upon to pay the toll of life and property," ran the editorial on August 14. The "vessel was manned by a splendid lot of men, types of those who man the fishing fleet that yearly sails from Lunenburg....[G]loom and sorrow is [sic] prevailing in every district in the presence of the dread tragedy, almost unprecedented in the history of the county's fishing industry."

But there was more bad news to come. On August 18 agent Harvey in Halifax received word from the Department of Marine and Fisheries steamer *Lady Laurier* that she had arrived at Sable Island on her regular trip with supplies and had found about twenty trawlers fishing in the vicinity. One vessel that should have been there, however, was not. Harvey instructed the wireless stations at Canso and at Chebucto Head to watch for signs of the LaHave schooner *Sadie A. Knickle*, which had not been heard from since the night of August 7. Some believed she had gone off to the Grand Banks and would return with the other members of the Lunenburg fleet, now nearing completion of their summer trip. This theory was encouraged by a report from Walter Blank, the man in charge of No. 3 Station on the island, a man who knew "the tempers and tempests of Sable Island." Blank said that on the early evening of August 7, just before the storm broke, he had seen the *Sadie Knickle* travelling under full sail and going east on the south side of the island, which could possibly mean she was headed towards the Grand Banks.

In fact, there were two LaHave schooners missing, the *Sadie A. Knickle* and the *Annie E. Conrad*, Captain Fred Richards. The latter turned up safely in a couple of days, and there was hope that the *Sadie Knickle* would soon follow. Even ten days after the storm, on August 17, Captain M. J. Parks, managing owner of the vessel, felt certain that all was well. The schooner was not expected in port yet. Likely she was fishing on Quero Bank, some sixty kilometres east of Sable, or she may have gone to fish on the Middle Ground, the area of the sea between the inner shoals and ridges and the banks.

Sadie A. Knickle

The first real evidence of disaster came on August 21 from the Lunenburg trawler *Marian Elizabeth*, Captain John Westhaver. On arrival at Canso the crew said that the trawler had passed through wreckage strewn over the ocean for some distance, including a deckhouse and a ship's knee. It couldn't belong to the *Sylvia Mosher* since her ruins were intact on the Sable Island bar. Later the same day Harvey received word from the steam trawler *Lemberg* of a wreck southwest of Sable, with two masts projecting six metres above the water, apparently attached to a submerged wreck.

At the suggestion of H. F. Zwicker of Lunenburg, the Canadian government fishery patrol steamer *Arleux*, Captain H. P. Cousins, proceeded to Sable Island to determine once and for all the fate of the *Sadie A. Knickle*. On August 26 the search party found a battered water tank washed ashore on the Northwest Bar, and farther out on the bar a splintered flour barrel with "Sadie Knickle...La Have" marked on the head in blue pencil. At last, said the *Halifax Morning Chronicle* of August 27, after almost three agonizing weeks of waiting, her fate was known. The *Sadie A. Knickle* had been swept onto the treacherous bar.

There was one man on board the *Arleux* to whom the loss of the crew meant as much "as his own life and happiness." Captain Parks, part owner of the vessel, "a big man with the hardy strength of the Lunenburg County race," had known the men intimately for a lifetime, and was familiar as well with the home life of their families. He was badly shaken when it was decided that further searching was useless. As the *Arleux* sailed away from Sable Island he stood by the rail, dry-eyed, grey of face, and gazed over the waves he had sailed over for so many years. He had hoped to identify the dories, but "smashed and bruised" they "lay like scarred relics along the long sandy beach, a mute warning of the crudeness of a sea which at the moment lay blue and smiling." He had immediately identified the flour barrel as being one of an order he had received from M. J. Ritcey of Halifax when the vessel had been outfitted for the season.[3]

On August 25, 1926, J. J. Kinley, former mayor of Lunenburg and member of the legislative assembly, commented in the *Progress-Enterprise* on the loss of the *Sylvia Mosher* and the growing concern for the *Sadie A. Knickle*. There should be some method of warning vessels fishing near Sable Island of an approaching storm, he said. Sometimes gales come up so fast even a falling glass doesn't give sufficient notice to vessels at sea. The weather bureaus broadcast warnings in advance; Sable Island must have had this warning by wireless. There could be prearranged signals to let ships know from the island that they should stay clear: a flare by night or gun by day to warn of impending danger. The *Canadian Fisherman* of September 1926 also argued that in light of the "stark tragedy," storm warnings were urgently needed. There should be some means of imparting the weather forecasts from Sable Island to nearby vessels, especially to the handliners who fish closer to shore. "Let us have no repetition of the disaster that has just occurred. The toll of the sea is heavy, and even one man is missed."

Lunenburg's William Duff, president of the *Progress-Enterprise*, though defeated in the federal election at home in Lunenburg County, had become in a by-election the Member of Parliament for Antigonish-Guysborough. He argued forcefully for the installation of transmitting and receiving radio equipment aboard all Lunenburg schooners. The fishing vessels, he told the Department of Marine and Fisheries at Ottawa, should receive weather forecasts and storm warnings by radio telephone. He also tried to persuade the department to provide sufficient money so each vessel could be equipped with a receiving set. According to the *Progress-Enterprise* of October 20, the department replied that it was taking the matter up with the Canadian Marconi Company with a view to equipping Louisbourg station with a high-powered radio telephone attachment to provide service twice a day.

The premier of Nova Scotia, Edgar Nelson Rhodes, when told of the loss of the *Sadie A. Knickle*, declared: "The whole province mourns the loss of this brave band of men, most of whom were taken from us in the full flower of their manhood.... [I]t should be our firm resolve to see that in future there is placed at the service of our fishermen every aid which modern scientific invention has produced to the end that all hazards shall be reduced to a minimum."[4]

William Lyon Mackenzie King, the federal Liberal leader, while campaigning in Yarmouth in late August for the September 14 election, in which he would defeat Arthur Meighen, extended his deep sympathy for those lost at sea. "Those engaged in fishing," he stated, "were not only engaged in the production of natural wealth but were the first line of defence in the event of an attack upon our shores from across the sea"—a rather odd statement given the nature of fishermen and their calling. From Yarmouth, according to the *Herald*, King made a "triumphal tour" eastward by train along the South Shore to Liverpool, the town "brilliant with bunting"; "our governments," he said, "must give more attention to the great basic industry of fishing." From Liverpool King motored to Lunenburg, arriving late that Saturday night. Met by a band at the entrance to town, he was swept to the centre by a long line of cars and a torchlight procession. The post office clock had struck midnight, but something had to be said to the crowd; amid much laughter, the Liberal chieftain declared that he did not desire to add anything more to "the Conservative scandal campaign" so he would refrain from breaking the Sabbath with political speeches. Earlier, in Liverpool, King had declared: "It is the duty of the State to see that everything should be done to secure the safety of men in the fishing industry."

Alas, no one hurried to do anything.

The August gale of 1926 had taken two vessels and the lives of many men. The owners suffered financially for the loss of the schooners, but it was the loss of the crew that mattered most. Twenty-five men died on board the *Sylvia Mosher*, and another twenty-five were lost when the *Sadie A. Knickle* went down.

The *Sylvia Mosher*, the first of the wrecks to be discovered, had been launched as recently as the spring of 1925 at a cost of $2,496 by the John McLean & Sons shipbuilding company in Mahone Bay. According to the *Halifax Herald* of October 2, 1926, there had been a hitch at the launching as the vessel went over on her side; she was righted, however, and a couple of days later slipped down the ways and launched herself.

At 71 tonnes, fitted with a sixty-horsepower Bergsunds crude oil engine as well as her sails, and with an eight-horsepower hoisting engine on deck, the *Sylvia Mosher* was considered up to date in every respect. Her outfitters, and part owners, were Robin, Jones & Whitman of Lunenburg. In December 1925 she had paid out to her owners for her season's work $12,064, which was nearly 50 per cent of the capital invested, or $189 a share, the biggest dividend of any vessel. The *Sylvia Mosher* caught 325,105 fish that year.

Captain John D. Mosher, Lunenburg

The captain of the *Sylvia Mosher* was John D. Mosher of Lunenburg, age thirty-three, the son of Alfred Mosher of nearby Corkum's Island. He was in the unique position of being highliner of the Lunenburg fleet for three consecutive years, a feat unequalled in the history of the fishing industry. In 1923 and 1924 he had been captain aboard the *Pauline E. Mosher*, but in 1925 he had the new schooner built. "A Money Making Skipper," ran the headline of an article about him in the *Canadian Fisherman* of March 1926. The *Sylvia Mosher* had been christened by his wife, Florence, pregnant at the time her husband was lost, and the vessel named for the captain's two-year-old daughter, Sylvia. In Lunenburg in 2013 Sylvia (Mosher) Cart said that her mother, after the loss of her husband, had been forced to give up their house, which was just being built; she moved in with her parents, and her two daughters were raised at their home. John Mosher was the eldest of six brothers, all of whom followed the call of the sea.

Several fishermen on schooners that survived had spoken to the crew of the *Sylvia Mosher* before the storm, and had learned her hold was full with a catch of 2,200 quintals of fish. They suspected the vessel was likely too heavily laden to survive once trapped by the storm. The fish caught that day were to be kept by the fishermen themselves for their own winter supply, and so they would have been happy to go on fishing, even if the sky looked threatening.

Among the twenty-five men on board was the captain's brother, Aubrey Mosher, twenty-three, from Corkum's Island, who had sailed on the voyage as second hand. As was often the case on the banks schooners, there were other members of the crew who had relatives aboard. There were, for instance, four men named Baker from the LaHave Islands, at the mouth of the LaHave River: Caleb Baker and his son Blanchard, Caleb's brother Guy, and Arthur Baker. At the time of the gale a man named Ernest Baker was trawling on the eastern side of the Grand Banks almost five hundred kilometres from Sable Island. Many years later, in an interview with Simon Watts in the *National Fisherman*, he said: "We knowed nothing about what was going on. Every week we used to sail into Newfoundland for bait and we was on our way in. That night we had no wind, but the sea was so great that we had to lower every stitch of canvas and just let the vessel lay. We knew there had to be an awful storm somewhere." When the captain returned from shore, having gone to inquire about bait, he called Baker to "come back aft" and told him that the *Sylvia Mosher* had been lost with all hands. Baker had two brothers aboard her, his oldest and his youngest. There were seven gone from the islands. "Knowed them all my life. It was a hard slap, I'm telling you." Captain Angus Walters, in the *Herald* of August 27, on returning from the Grand Banks, confirmed that the schooners there did not experience the storm that had wrecked the vessels off Sable Island.

Also on board the *Sylvia Mosher* was a man named Enos Baker, age twenty-seven, from Dublin Shore, in the western part of Lunenburg County near the LaHave Islands. A fisherman who had survived what he called the "August Breeze" ("An' there was seas I'm safe to say a hundred feet high"), sailing aboard the *Silver Thread* with Captain Murdock Getson, mentioned the Bakers in his account of the storm in Peter Barss's *Portrait of Lunenburg County*:

> [I]t's strange how t'ings happen.... Titus Baker from West Dublin an' Enos Baker—that was his son—they always went toget'er aboard of a vessel. They always slept toget'er in a double bunk. An' this year they separated an' Enos went wit' Cap'n John Mosher in the *Sylvia Mosher*, an' Titus, he went wit' us in the *Silver Thread*. We was to the west'ard o' the *Sylvia Mosher* an' we never seen 'er when we was out. But when we got in we heard she was lost an' all hands gone. Didn't know nothin' about it till we come home...an' they didn't return. An' when Titus heard it...he almost went crazy. Boys, he felt some bad.

John Eldon Wagner, Stonehurst, Lunenburg County.

There was grieving as well farther down the shore, in the village of Vogler's Cove. Bertha Whynacht of that village, wrote Clara Dennis in the *Halifax Herald* of October 2, was forced to mourn her three sons, Kenneth, Wildon, and Ladonia, as well as her grandson, Kenneth's son Donald. Young Donnie was lost on his sixteenth birthday, on his second trip to sea. It was the first time he'd sailed with his father. Ladonia, who was living in Liverpool at the time, was the cook aboard. Kenneth Whynacht had gone to sea at the age of twelve. He often told his wife not to worry, that he was as safe on the water as on land. He'd worked in the United States in 1925, but had returned home to fish because he didn't want to work on Sundays. Also on the vessel was Moyle L. Whynacht from First South, a community near Lunenburg. Bertha Whynacht had three more sons fishing out of Gloucester. She was well familiar with the terrors of the sea for her husband was also a fisherman. Many years earlier his vessel had been wrecked off the Newfoundland coast and given up as lost. All through the fall and winter Bertha mourned, but with the coming of spring her husband returned. The vessel had been wrecked but the men had been cast ashore in an "unhandy" place and been obliged to rough it there throughout the winter.

On board the *Sylvia Mosher* were five men from the seaward end of the Lunenburg Peninsula. Leaman Graham and Rounsfell Greek were from the village of Blue Rocks. A tombstone in the Upper Blue Rocks Cemetery reveals that Rounsfell Greek's wife Clydie lived until 1998, seventy-two years after her husband drowned. James Tanner and two young brothers, Warren Wagner and John Eldon Wagner, came from nearby Stonehurst; according to Betty Fralick, a family member, the sister and niece of Warren and John blamed themselves for the loss because they'd been playing cards on Sunday while the men were out to sea. Calvin Tanner of First South was only fifteen years old when he drowned.

Rounsfell Greek, Blue Rocks, Lunenburg County.

Hasting Himmelman (centre), Bridgewater, and fellow fishermen.

Frank Emmanuel Walfield, from the LaHave Islands, had been in the Nova Scotia Regiment in the First World War. Like many others he had been drafted under the Military Service Act 1917, and was signed up at Bridgewater on October 26, 1918, towards the end of the war. On the military form, which he signed with an X, he gave his religion as Methodist and his trade or calling as "fisherman."

Finally, one might note Hastings Himmelman from Bridgewater. This was to be his last trip, for he planned to be off to the United States to find work when he returned. On the night of the gale, Hastings's wife Nita was writing a letter to her husband, and her daughter Mildred, age seven, had written a note to be included with it. The letter was never sent. Sometime later, when Nita returned home from picking raspberries, she found the minister waiting for her with the news.

The story of the sinking of the *Sylvia Mosher* made its way into literature. In the novel *Rockbound*, Frank Parker Day's hero, David Jung, an inshore fisherman from Rockbound (based on the island of Ironbound in Mahone Bay), gets a chance to sail on a banks schooner. In the novel[5] Day imagines the *Sylvia Mosher*'s end:

> At that moment the *Sylvia Westner* struck. Hand-line Johnny had no luck that night. All was over in the twinkling of an eye. The vessel, deep-laden, was travelling at the rate of twenty knots, and a tooth of black bottom rock whipped bottom and

keelson from her as cleanly as a boy with a sharp jackknife slits a shaving from a pine stick. Two thousand quintals of split fish and the unwetted salt dropped down upon the yellow sands.... Within ten seconds of her striking, every man of the crew was in the sea...poor scraps of humanity, weighed down with soaked clothing and long boots; a flash of yellow oilskins, hoarse cries that made no sound in the fierce tumult, and they were gone. Some swam a stroke or two, some clung for an instant to trailing rigging or broken dory, but few clung long in that mad breaking sea.

Day's fictional hero lashed himself to a dory's painter and was washed ashore, but on the real *Sylvia Mosher* there were no survivors.

On November 24 a "pathetic relic of a great sea tragedy" was reported in the *Progress-Enterprise*. A small wooden box belonging to Freeman Corkum of Feltzen South, who had sailed on the *Sylvia Mosher*, was picked up on the beach at Sable Island not far from the wreck. It was sent to C. H. Harvey at Halifax by Superintendent Henry and identified from papers as Corkum's property. There were scissors, a jackknife, a memorandum book, wool for darning socks, and other odds and ends. The box and its contents were forwarded to the lost fisherman's widow.

There were, apparently, sightings of the *Sylvia Mosher* herself. Bruce Armstrong, in his book about Nova Scotia's mysterious island of sand, told of strange happenings: "Sable was a place where spirits haunted the living and even ships returned as apparitions." Not long after the disasters, he wrote, the keeper of the East Light complained of seeing visions of the *Sylvia Mosher*. "The apparition would appear through a glowing phosphorescent haze; the ship would seem suspended in space, hanging above the ocean. Ghostly figures of the crew could be seen in motion on the decks. They were in panic as winds gripped and shook the phantom vessel." The keeper watched the "terrible spectacle" as "a steady trickle of men leaped over the sides, one after the other, to vanish into the raging sea below. Their bodies made no splash as they struck water and like a silent film of that era, the action was soundless, adding to the horror of the vision." Not surprisingly, the lighthouse keeper begged to be taken off the island.

> ### LOST ON THE *SYLVIA MOSHER*, AUGUST 7–8, 1926
>
> **Captain John D. Mosher**, 33, Lunenburg
> **Aubrey Mosher**, 23, Corkum's Island, brother of John D. Mosher
> **Freeman Corkum**, Feltzen South
> **Calvin G. Tanner**, 15, First South
> **Leaman Graham**, 20, Blue Rocks
> **Ephriam Rounsfell Greek**, 32, Blue Rocks
> **James Tanner**, Stonehurst
> **Warren Wagner**, 22, Stonehurst
> **John Eldon Wagner**, 19, Stonehurst, brother of Warren Wagner
> **Caleb Baker**, 41, LaHave Islands
> **Blanchard Baker**, 16, LaHave Islands, son of Caleb Baker
> **Guy Baker**, 19, LaHave Islands, brother of Caleb Baker
>
> **Arthur W. Baker**, 25, LaHave Islands
> **Enos Baker**, 27, Dublin Shore
> **John Edwin Bell**, 25, LaHave Islands
> **Fred Cleversey**, 28, LaHave
> **Melvin Richards**, 20, LaHave Islands
> **Frank Walfield**, 32, LaHave Islands
> **Kenneth A. Whynacht**, 45, Vogler's Cove
> **Donald R. Whynacht**, 16, Vogler's Cove
> **Wildon Whynacht**, 42, Vogler's Cove, brother of Kenneth Whynacht
> **Ladonia Whynacht**, 35, Liverpool, brother of Kenneth Whynacht
> **Moyle L. Whynacht**, 33, First South
> **Hastings Himmelman**, 34, Bridgewater, originally from LaHave Islands
> **Adam Selig**, Lunenburg
>
> All of the men were from Lunenburg County, although Ladonia Whynacht seems to have been living in Liverpool, Queens County, at the time.

In the fall of 1926, on October 7, the *Progress-Enterprise* announced that a new handliner had been launched from the yards at John McLean & Sons in Mahone Bay. Designed by C. A. McLean, the *Robert Esdale* was built for Captain Allan Mosher, the late John Mosher's brother. A sister ship to the *Sylvia Mosher*, she would sail to the banks in the spring with the Lunenburg fleet. Hope springs eternal! Nova Scotia folklorist Helen Creighton in the late 1920s spent an afternoon on a vessel captained by a brother of the skipper of the ill-fated *Sylvia Mosher*. He too fishes off Sable Island, she wrote in *Maclean's* on December 1, 1931, "and in his blue eyes there is something that tells that he is defying the fates. It is a war waged with Sable, which he hates for the brother it has taken from him."

The LaHave Islands, southwest of Lunenburg at the mouth of the LaHave River, and several other small communities west of the river were particularly hard hit by the August gale of 1926. As well as the thirteen men from that area of Lunenburg County who died on the *Sylvia Mosher*, another thirteen died on the *Sadie A. Knickle*.

Unlike the *Sylvia Mosher*, the *Sadie A. Knickle* was not a new boat; she had been built in 1918 by the Nova Scotia Shipbuilding and Transportation Company at Liverpool for Captain Roland Knickle of Lunenburg. At 88 tonnes and 34.5 metres long, she was slightly larger than the *Sylvia Mosher*. She was now owned by as many as forty shareholders scattered widely throughout the county, including Captain M. J. Parks, Captain Charles J. Corkum, and the Lunenburg Outfitting Company operated by William Duff. On this her last trip she had sailed from LaHave, her home port, around the first of July. Captain Corkum of Mount Pleasant, fifty-four years old, left behind four children. He first went to sea at the age of nine, and had followed the sea for forty-five years.

Captain Charles Corkum, Mount Pleasant, Lunenburg County.

Whereas the crew of the *Sylvia Mosher* was made up of natives of Lunenburg County, the men on the *Sadie A. Knickle* belonged to widely separate parts of Nova Scotia. Four men came from as far away as Chéticamp, on the northwest coast of Cape Breton. Joseph Chiasson, age twenty-nine, had been married for only three years; his youngest daughter, Marguerite-Thérèse, had been born only a few months earlier, and his eldest daughter, Marie-Luce, died that same year. Amédée Chiasson was only twenty-five and single at the time of the shipwreck. Stanislaus Muise, in his mid-forties, had married Marie Aucoin in Passchendaele, part of Glace Bay, in 1910; he was a descendant of Philippe Mius d'Entremont, Baron de Pobomcoup (Pubnico), who came to Nova Scotia in the 1600s as commander of the king's troops and stayed on as seigneur over a large piece of land near present-day Yarmouth. The fourth man, Cyrille Chiasson, age thirty-four, was Joseph's older brother, and he was married to Stanislaus Muise's sister.[6] The men from Chéticamp came from a fishing community with a long and distinguished history, according to historian Ruth Fulton Grant. A man named Philip Robin had come out from Jersey about 1764 to establish a fishery, made the town his headquarters, and marketed the cured product in Europe. Chéticamp remained the base for a successful fishery for over a century.

Also aboard the LaHave handliner were five men from small communities down the shore in Shelburne County. One of them

Stanislaus Muise, Chéticamp, Inverness County, Cape Breton.

was Samuel Firth, sixty-five, a widower from Sandy Point with two children; his wife had died age thirty-six and because Samuel was so often away at sea his children had been brought up by their grandmother. Ross Pierce, from Lower Sandy Point, who sailed with his brother-in-law Horace Rhyno, left behind a son named Ross; born on August 3, 1926, only a few days before the *Sadie A. Knickle* foundered, the child died young from tuberculosis. One of the older men aboard from Shelburne County, John Baptiste, as a young man had survived the wreck of the Lockeport schooner *Mary E. Harlow* in February 1895 when she ran aground off Port Joli Beach on the way to the Turks and Caicos Islands.

Thomas Martell of Halifax, formerly of L'Ardoise, Cape Breton, came aboard as cook. Unable to find work on shore he had joined the doomed crew. He had not gone to sea for over twenty-five years. On August 30 Martell's body washed ashore on the Northwest Bar of Sable Island. A description of his features and clothes was relayed by wireless to his daughter Marie, who identified him. On his last visit home he had asked Marie for a crucifix given to her by her grandmother to take with him on the trip. His wife dead, he left behind six children, the youngest a boy of seven. Thomas Martell and one unidentified man, whose body was badly decomposed and "attired in pajamas," were apparently the only bodies to be recovered. Both men are buried on Sable Island.

Walter Wamback, Mount Pleasant, Lunenburg County.

But over half of the crew, including the captain, came from the western part of Lunenburg County. There were four Wambacks: Walter from Mount Pleasant, William from Broad Cove, and Wade and Parker, teenage brothers from New Cumberland near

Mount Pleasant. Walter Wamback had named his son Clayton Walter Wamback after the schooner *Clayton W. Walters*, in which he had shares, but because the *Sadie A. Knickle* was closer to home he had sold those shares and sailed on her instead. Captain M. J. Parks arrived in a Model "T" Ford, wearing a black armband, to tell Walter's wife that the vessel had been lost. The family had waited nineteen days for the sad news to be confirmed.

Andrew M. Shankle of Pleasantville died on the *Sadie A. Knickle* along with his son Basil; in his early teens, Basil was on his first trip to sea and he sailed as a flunkey. From the LaHave Islands came Simon T. A. Bush, his brother Robert, and Robert's son Redvis. Some fifty years later, in Peter Barss's *A Portrait of Lunenburg County*, Simon Bush's son reflected on the loss of the *Sadie A. Knickle*: "My father was into 'er, an' his brother, Robert. An' that was the first trip that Uncle Robert had his son wit'…Redwiss…he was only young…fourteen he was. Well, there was t'ree right from one house.…My father an' all them men…gone.…That storm killed the islands. It was just the same as somethin' come right down o'er the islands, you. You see, it was all young men that was drowned…all of 'em. Nobody felt like havin' dances or anythin' else wit' all our young people gone…we had no courage for anythin' like that. Everybody felt too bad about it. An' it…honest…it was never the same out there."

A rather curious story emerged about the sinking of the *Sadie A. Knickle*. One Rev. D. M. Matheson, while out walking on King's Head Beach, Pictou County, claimed to have come upon a bottle with a note inside. The message read: "Help. All hope is abandoned. Are out in life boat. Capt of Sadie Knickle." If true, the bottle had travelled a very long way, over three hundred kilometres, from Sable Island to the coast of the mainland, through the Strait of Canso, and up the Northumberland Strait to the beach. The *Halifax Herald* reporter who told of the incident on September 3 sounded sceptical.

Two days after the *Sadie A. Knickle* was wrecked, according to Helen Creighton, a man named Gregoire, on the government staff, swore that he saw the vessel with the men hoisting sails. Ghosts as well as horses live on Sable Island.

LOST ON THE *SADIE A. KNICKLE*, AUGUST 7–8, 1926

FROM LUNENBURG COUNTY:
Captain Charles J. Corkum, 54, Mount Pleasant
Simon T. A. Bush, 42, LaHave Islands
Robert Bush, 39, LaHave Islands, brother of Simon Bush
Redvis Bush, 14, LaHave Islands, son of Robert Bush
Norman Conrad, Cherry Hill
Perry Corkum, 39, Pentz
Robert Haughn, 33, Pleasantville
Andrew M. Shankle, 56, Pleasantville
Basil C. Shankle, early teens, Pleasantville, son of Andrew Shankle
Walter Wamback, 53, Mount Pleasant
William Wamback, 40, Broad Cove
Wade Wamback, 19, New Cumberland
Parker Wamback, 17, New Cumberland, brother of Wade Wamback

FROM SHELBURNE COUNTY:
John Howland Baptiste, 55, Jordan Ferry
Burns Buchanan, 24, Lower Sandy Point
Samuel A. Firth, 65, Sandy Point
Hugh Dunlop Moulton Pierce, 68, father of Ross Pierce
Ross Pierce, Lower Sandy Point
Horace Rhyno, Sandy Point, brother-in-law of Ross Pierce

FROM INVERNESS COUNTY:
Cyrille Chiasson, 34, Chéticamp
Joseph Chiasson, 29, Chéticamp, brother of Cyrille Chiasson
Amédée Chiasson, 25, Chéticamp
Stanislaus Muise, mid-40s, Chéticamp, brother-in-law of Cyrille Chiasson

FROM RICHMOND COUNTY:
Amos Burke, 20, St. Peter's

FROM HALIFAX COUNTY:
Thomas Martell, 53, Halifax

The *Shelburne Gazette and Coast Guard*, August 26 and September 2, 1926, claimed that a man named Henry Hemeon of Sandy Point, a widower, was on board, but his name is not on the memorial in Shelburne. Hugh Pierce, on the other hand, is not mentioned in the newspaper but is on the memorial. It would appear that Ross Pierce's wife, Alma, lost not only her husband and brother, Horace Rhyno, but her father-in-law as well.

❧ ❧ ❧

People asked why good men had to die, to which there was no answer. Had there been no warning of an approaching storm? On Friday, August 6, the weather forecast on the front page of the *Halifax Herald* stated: "Moderate winds; fair. Probably followed by showers by Saturday." The next day—the storm would hit that evening—the forecast read: "Increasing winds; probably gales; cloudy at first; followed by rain. Tropical storm centered tonight north of Bermuda, southbound mariners should use caution." Also on the front page of the *Herald* on August 7, in heavier print: "Storm Warning. Washington, Aug. 6. The Weather Bureau tonight issued the following storm warning advisory, 9.30 P.M. Tropical storm of hurricane intensity central tonight about latitude 35, longitude 64, moving north-north-east toward North Atlantic steamer route. Extreme caution advised vessels north and northeast of center."

Of course none of this meant anything to the vessels fishing off Sable Island, for they had no radios.

People also naturally wondered how the men died. Did they take to their dories and abandon the vessel, or were they swept overboard one by one into the terrible seas as the schooner was beaten to destruction? Or were they aboard the vessel when it was driven in upon the shoals, as had happened to the *Sylvia Mosher*, burying them in the ever-shifting sands of the graveyard of the Atlantic? As Clara Dennis wrote in the *Herald* on October 2, about Captain Corkum's wife in Mount Pleasant, "Not alone for her loss does she sorrow, but for the kind of death he died; what he went through beforehand; the final terribleness of drowning. The desolation of never looking on his face again! Denied the consolation of those who minister to loved ones in their dying moments. Denied even the consolation of a grave."

There was much speculation about what might have happened to the vessels. The *Canadian Fishermen* suggested that the *Sylvia Mosher*, for instance, had either capsized and was later thrown on the bar when the wind shifted or else she had been caught too close to shore and driven onto the treacherous sands. A writer in the *Herald* on August 19 speculated, rather vividly, on the last moments of the doomed vessel. The "once proud fisherman of the Lunenburg fleet," wrote J. J. T., was now but "a shattered hulk…a battered derelict." He suggested that the captain had decided to run with a light canvas before the wind towards the western end of the island:

> It would only require a flying jib sail to send the craft bowling along at a fast rate before such a wind. The island was rapidly approaching. West Light was showing up on the port bow. Stygian blackness and only the flashing hope of the sailor's friend as a guide. Hauling off before such a wind might send her into the vortex of raging breakers ahead, broadside on to certain destruction. She had crossed this western bar before. There were numerous channels. It was high water. Anyway to turn back now was impossible. Through the hissing foam-lashed breakers she must go—and through the other side to safety and shelter in the lee of the island on the northern side. But the doomed vessel did not win through.
>
> With the hurricane driving her sheer hull through the spume-drift and tempest-lashed shallow water over the bar at a speed calculated to be anything between twelve to sixteen knots, she grounded on a shoal, her whole bottom was ripped asunder in one fell sweep and she turned over and over, her masts wrenched completely from their holdings in the

keelson and everything movable, living and inanimate gone in one catastrophic crash that could have taken but moments to happen.

More prosaically, and perhaps more reliably, Reuben Naugle, coxswain of the life-saving crew on Sable Island, had his own ideas about the fate of the *Sylvia Mosher*. Naugle went aboard "the pride of the Lunenburg fleet" as she laid breaking to pieces beneath the cruel pounding of the sea, and attempted to "reconstruct something of the gallant fight for life which ended in failure." He discovered that the wheel had been lashed, which suggested to him that the schooner could no longer be guided by the men, who likely took to the rigging. There were indications that the vessel had broken clear from her anchor and therefore drifted helplessly before the winds. Her sails were torn to ribbons and the foremast was lying intact alongside the wreck, which probably showed that the vessel had struck very hard with sufficient strength to dislodge the foremast. The amount of sand on deck suggested that the ship had struck the Northwest Bar before drifting to her final resting place on the sandy shoals off No. 4 Station. "Drifting before the winds," Naugle concluded, "her anchor gone, and with thunderous seas breaking over her, the *Mosher* was helpless." Were the men swept to their deaths from the rigging one by one, or were they jolted from their temporary safety by the jar when the ship struck the bar? No one will ever know for sure what happened.[7]

Many months later, on January 4, 1928, Captain Gjert Myhre of the trawler *Venosta* claimed in the *Herald* that he had seen the *Sylvia Mosher* an hour before the great August gale struck. Sixteen dories were strung out around her. "The fog closed in," he said, "and with it came the wind. I don't believe a man ever got back to that vessel."

Some small idea of what the men went through can be gathered from accounts of the storm told by survivors. Captain Harris Conrad of the fishing schooner *Mary Ruth* arrived back in port a few days after the gale with a broken shoulder blade. Nine men in his crew were seriously injured, reported the *Bridgewater Bulletin* of August 17, but they had escaped the wrath of the storm. The schooner was "literally the toy of the tempest for all of Saturday night and part of Sunday," Captain Conrad said. The storm hurled the vessel onto the Northwest Bar of Sable Island, and the surf swept her clear of everything left on deck, sails torn, all dories gone, even the cabin stove smashed to pieces. She was lifted as by an unseen hand and hurled free of the sandbar into the clear water on the other side.

Two other vessels, the *Golden West* and the *Silver Thread*, were saved by the desperate means of "crossing the bar": sailing over Sable Island's sandbars into the relative safety of the deeper water beyond. One young fellow, age thirteen, who went "vessel fishin'" as a header that year with Captain James Getson on the *Golden West*, later recalled the storm in Peter Barss's *Portrait of Lunenburg County*: "Oh my, oh my, the sea...was vicious. An' blowin'. It was wicked to the world—blowin' you. That was the biggest storm ever I was into." The skipper's brother, second hand on board, came forward to the forecastle at the height of the storm and spoke to the men. "You got any friends home now," he said, "it's time to t'ink on 'em....Get ready. Everybody stays on deck....I t'ink we're all hands gone. If she strikes the bar, every man for himself, the Devil for us all." Some of the men stayed in their bunks, reckoning that if they were going to die, it might as well be there as on deck.

Percy Amos Baker, from the LaHave Islands, was also on the *Golden West* when she crossed over the bar. Late in life Baker described the storm as it approached: "[A] suffocating stillness with a glassy sea coming in giant rollers. A dark sky standing almost at the masthead. Then all hell let loose with no time for the sea to make breaking waves, rather giant gusts which dashed the sea gulls flat on the water to drift away dead." With the anchor dragging and the decks awash—"half the crew down below praying, half at the pumps"—Baker told a fellow crew member to grab a rope, tethered himself to the other end, crawled forward and cut the anchor cable with an axe. The bow of the ship rose right up in the air, then down with the stern almost vertical, but the torrent of water breaking over the deck stopped. The vessel was saved and Percy lived to be ninety-eight. His nephew Caleb and Caleb's sixteen-year-old son Blanchard died on the *Sylvia Mosher*.[8]

Did experienced seamen not have even an inkling of a coming storm? Herbert Getson, on board the *Golden West*, claimed that one of the old sea dogs said it would be "a summer breeze." Another captain's story was related by Fred Winsor in his article "Solving a Problem": "That's what you call a fisherman's luck. You're anchored in the middle of the ocean and you got to take it as it comes....Well the weather-glass gave no warning. It showed nothing at all. At nine o'clock there was an ordinary breeze but at ten o'clock the sea come ahead of the wind and you knew there was somethin' back of it driving it....We had twenty-four hundred quintals of fish when the gale come and the vessel was like a log. At twelve o'clock she broke adrift...."

In an item headlined "All Canada Mourns," bordered in black, the *Halifax Herald* on August 27, 1926, declared the Nova Scotia fishermen to be a special breed of men:

> Never again will the topsails of those two staunch schooners appear over the headlands, heralding their homecoming with bounteous fares....There is a great grief in many homes in this old Province today, but a grief sweetened by the thought that those who have gone—the breadwinners of the households—gave up their lives in a hazardous calling that is of the essence of Nova Scotia existence.
>
> We bring no store of ingots, no spice or precious stones,
> But what we have we gathered by sweat and aching bones.

That is the experience of the men who go down to the sea in ships to reap the harvest of waters. It is a hard trade and filled with struggle. Sometimes the sea conquers....It needs a rugged race and a brave race to battle the sea and its terrors. Such is the race of men bred in Nova Scotia.

There were, of course, memorial services, to honour and to remember the men who had died. On September 19, at the United Church in Vogler's Cove, a memorial service was held for the Whynacht brothers and Kenneth's son Donald, all lost on the *Sylvia Mosher*. "Their untimely but honorable exit," wrote a reporter in the *Progress-Enterprise* on September 22, "was met, like that of many former sea heroes, while engaged in the 'front line' of Canada's most ancient industry, winning sustenance for their families and perpetuity for society and the state." A memorial service was also held for Enos Baker at St. John's Episcopal Church at West Dublin.

On September 28 the *Bridgewater Bulletin* reported on a service held on Sunday eight days earlier at Zion Lutheran Church in Lunenburg to pay respects to Captain John Mosher and his brother Aubrey. It was estimated that a thousand people filled the church, and many others could not get in. The local Orange Lodge attended the service in a body to pay respects to their brother, John Mosher. Three memorial gifts were presented, one planned by Captain Mosher himself as a memorial to his mother: a set of auditorium chimes, electrically operated from the console of the organ.

The *Lunenburg Argus*, on October 7, observed that Nova Scotians paid tribute to their soldiers, their pioneer settlers. Why not their fishermen? Various mining communities had erected monuments to the memory of men who had lost their lives in mine explosions. Two weeks later, on the twenty-sixth, the *Argus* raised the issue again, claiming it was well to postpone the annual fishermen's picnic this year because too many fishermen's homes had been saddened by recent marine catastrophes to attempt a celebration. Let the moneys on hand be used, the paper said, towards erecting a monument for those in this county who have lost their lives on the "briney deep."

A memorial service for the men of Mount Pleasant who died on the *Sadie A. Knickle* was held in the village on September 12, 1926. So many people attended that the church's organ had to be removed to the outer steps, and chairs for the mourners were arranged on the grass near the door. About a thousand people, flanked by members of the Oddfellows Lodge, assembled on the green slope for the open-air service. In April of the following year an impressive service was held at the United Church. A tablet designed by the pastor, the Rev. N. Cole, was unveiled by two little girls in white dresses, Vivian and Rubie Wamback. The memorial was in the form of a large lifebuoy, in relief, upon white marble, and the inner circle ground bore the names of the four men from that church who had been lost: Captain Charles Corkum, Walter, Wade, and Parker Wamback. Underneath were words taken from the Gospel of St. Mark, 6:48: "And the fourth watch of the night (Jesus) cometh unto them, walking upon the sea." During his five years ministry at LaHave, the minister noted, it had been his duty to break sad news of sea disasters to more than twenty homes.[9]

Announcing the large memorial service that would take place in Lunenburg on October 3, the *Argus* on September 30, 1926, declared: "Lunenburg County people have had many marine losses, but it is doubtful if they have ever been called upon to mourn the passing of so many of their fine men as they have this year." The *Halifax Herald* on September 1 said that the sinking of the two vessels with all hands was the worst disaster to befall the Nova Scotia fishing fleet since 1889, almost forty years earlier, when the *Morris Wilson* of Lunenburg and the *Georgina* of Yarmouth had been wrecked in a big storm off Sambro with the loss of both crews. The August gale of 1926, then, came as a tremendous shock to the people of Lunenburg County, because nothing comparable had happened in recent memory.

Surely 1927 would be a better year.

the AUGUST GALE of 1927

> But until an earthquake sinks Sable Island a thousand fathoms deep, the price of Atlantic fish will be the lives of men.
>
> —C. H. J. Snider, *Canadian Magazine* (1928)

DESPITE THE LOSS OF TWO SCHOONERS IN THE NORTH ATLANTIC GALES, 1926 was considered "a banner year" for the Lunenburg banks fishing fleet. The total catch of fish landed, as well as the average catch per vessel, had beaten all previous records in the history of the industry. Over the season a total of ninety-two vessels had caught 342,730 quintals of fish, each vessel catching an average of 3,725 quintals. The fleet was also expanding, for there had been seventy-six vessels in 1925 and only sixty-four in 1924. There was, then, cause for celebration. But unfortunately 1926 was, at the same time, not the most prosperous of years, for the price of fish was low, as much as two dollars a quintal lower than in the previous year. Financially, the *Canadian Fisherman* pointed out in November, the fishermen had not been very successful. The owners, too, had made little in the way of profits. A number of vessels had been so badly battered by the storms, with so much gear lost, that they would hardly be able to meet their expenses. In a year-end editorial on December 30, the *Lunenburg Argus* summed up the 1926 season by noting the low prices, which "must of necessity mean littler money for the fishermen." This shortfall was overcome to some extent, however, by "the remarkably large catch of the year." Lunenburg County had been particularly hard hit by the loss of life in the "appalling tragedy" of last August. "Let us hope," the newspaper continued, "that 1927 will see a larger catch and a higher price."

With plenty of provisions aboard to feed the hungry men of the crews, their holds filled with salt and bait, a few vessels left for the banks on March 9. The majority sailed on the fifteenth. Some of the first vessels to get away, the *Canadian Fisherman* noted in its April 1927 issue, were the *Partana*, Captain Guy Tanner, the *Mahala*, Captain Warren Knickle, and the *Grace D. Boehner*, under Captain Angus Tanner. These skippers were all young, the journal said, and their early start showed them to be "hustlers."

But trouble arrived early. In a fierce storm on April 7 and 8, the *Canadian Fisherman* reported in May, a gigantic wave boarded the schooner *Alsatian*, broke about four and a half metres up the jumbo stay, and came down directly on top of the crew of seventeen men. Miraculously, no one was washed overboard. However, crew member Robert Corkum of Pentz Settlement was thrown against the dory cradles and died almost instantly.

The first two trips in 1927 were not successful, for the weather was rough and the catch low. Discouraged by the low prices in the previous year, fewer fishermen had set out. The *Bridgewater Bulletin* of July 19 reported that fifty-eight vessels had gone on the frozen baiting trip, and only seventy on the spring trip (sixty-one trawlers and nine handliners), compared to eighty-seven the previous year. On the second trip a number of vessels had been caught in the ice in Sydney harbour, losing about two weeks of fishing. The catch turned out to be 30 per cent less than on the same trips in 1926. "This situation generally follows a season of low-priced fish," said the *Canadian Fisherman* in August, "as the people of Lunenburg County, who are very conservative about investing their money, will not invest their capital when the prospects for fishing are not bright."

There had been one promising development in 1926, namely the organization of Lunenburg Sea Products and the building in the summer of their new cold storage plant on land purchased from W. C. Smith & Company. Eventually the fishing vessels would be

Lunenburg waterfront.

able to obtain bait in their home port, rather than having it shipped in, and more money would stay in the county. There were plans, too, to carry on a large fresh fish and smoked fish business. But as the plant was not yet completed, it was not able to supply the whole fleet with frozen bait for the first trip in 1927.

On the spring trip the fishing vessels had spent four to six weeks on the North Atlantic. They were home now for a week or so, landing their fish and fitting out again. In early June they returned to the banks.

Toronto, Wednesday, August 24, 1927
Maritime Provinces: Winds; increasing to gales with heavy rain Thursday. Westerly winds; clearing. Tropical disturbance of moderate intensity centred south of Nantucket likely to move northward across Bay of Fundy and probably increasing in intensity.

The weather forecast for August 24, 1927, issued from the Dominion of Canada Meteorological Service and published by authority of the Department of Marine and Fisheries, spoke blithely of "moderate" winds, though there was a suggestion that they might become more intense. The next day the service reported "moderate westerly winds...The high pressure area now covers the eastern portion of the continent while the tropical disturbance has passed northeastward across the Maritime Provinces causing gales and heavy rain. The weather is now fair throughout the Dominion." Somehow, this doesn't quite capture the drama of the event.

There had been warnings that a storm was coming. The weather bureau in Washington on August 23 had reported a tropical disturbance east of the Bahamas that was "recurving to the northward...of great intensity and attended by hurricane winds at its center." A tropical storm was forecast—once again "of moderate intensity"—centred south of Nantucket and likely moving northward across the Bay of Fundy and probably increasing in intensity.

But none of this mattered to the fishing schooners already at sea on the banks. With no radio receivers, even an entirely accurate marine forecast would not have reached them. It was late in the season and the vessels were heavily laden with fish. Lacking communications equipment, they had no idea what lay in store for them.

Headed their way was a hurricane, the first tropical disturbance of the season, a storm that likely developed near the Cape Verde islands. It was detected some 480 kilometres northeast of St. Kitts in the Lesser Antilles on the morning of August 21. As such storms often do, it curved northward, making landfall only when it reached southwestern Nova Scotia as a Category 2 hurricane. The storm passed close to Nantucket during the daylight hours of August 24, reaching the Strait of Belle Isle between Newfoundland and Labrador by the morning of the twenty-fifth. It's hard to be precise about just what overtook the men and schooners on the banks, but it was clearly either a comparatively weak hurricane or a very strong extratropical storm, and it covered a vast area.[1]

On August 20, four days before the gale struck, H. F. Henry, superintendent of the life-saving station on Sable Island, was in Halifax for a bit of a break. The year had been "without untoward incident thus far," he said, no marine disasters. Indeed, the last wreck on the island had been the *Sylvia Mosher* in early August the previous year. Henry spent several days in the city, according to the *Halifax Morning Chronicle* of August 31, before returning to "his lonely little kingdom in the mid-Atlantic."

The "Great Storm"—referred to by the *Bridgewater Bulletin* on August 30 as the "so-called August gale"—swept the province from end to end in the early hours of Wednesday evening, August 24. Property damage, the newspaper estimated, was over a million dollars. Nine people around the province had been killed. Highways were gutted, bridges and railway lines washed out. Canoes were used to transport stalled motorists caught in floods. Telephone and telegraph lines were crippled. The town lights were out of commission, said the *Bulletin*, and all wire leading out of town was down. Bridgewater was cut off from the outside world! As were many other towns and villages, as well as Halifax, where torrential rains combined with "the terrific blast" to all but cut off the entire transportation and communication systems. The *Morning Chronicle* reported on August 25 that the storm had descended on the city with "the suddenness of one of Jove's thunderbolts." A travelling circus parked on the Common was in ruins, and a huge steel coal crane at Pier 9, worth thousands of dollars, had been blown down with a great crash; now it lay in "twisted impotence" on the piles of coal.

Damage was extensive around the province. Vessels were tossed about in Liverpool harbour, and many shade trees were blown down in the town. In Lunenburg the tide was the highest in years, and there was a great deal of damage to property and gardens. At Digby there were broken bridges, gaping holes in the streets; huge timbers that had broken loose from the piers and abutments were tossed about like matchwood, littering roads and impeding traffic. Yarmouth experienced the most violent storm since August of 1917, the heaviest rainfall in a seven-hour period ever recorded. The storm in Sydney was said to be the worst since the terrible storm of 1873.

Thousands of barrels of apples were blown off trees in the Annapolis Valley. Crops were "crushed to earth by the hand of an invisible reaper," laying barren fields that "but a short time before had been waving lanes of grain and corn." Silos were blown over at Middle Stewiacke. Two valuable horses were killed when a barn collapsed in Cumberland County. At Elderbank, in the Musquodoboit Valley northeast of Halifax, a farmer named Henry Killen had a "strange experience." While driving his cow to the barn a sudden powerful gust of wind lifted the animal off her feet and knocked her to the ground. Killen was dashed against the barn but escaped injury. Though it's not recorded, one assumes that the cow recovered from her undignified tumble.

Naturally there was great concern for the men and vessels out on the water. A man named Frank da Sylva from Blue Rocks and his fourteen-year-old son were caught by the gale about twenty-eight kilometres offshore. They started to run before the wind but in the darkness they were unable to make harbour. Dropping anchor near an island they endeavoured to ride out the storm rather than risk being cast ashore on the rocks. The small craft was pitched about like a straw by the huge seas; wave after wave broke over the men, but somehow, by baling and pumping, they survived, exhausted and thankful. Similar harrowing experiences faced inshore fishermen all along the coast. Caught in heavy seas on the fishing grounds, Warren Ossinger and his son Kenneth, from Tiverton on the Digby Neck, were less fortunate; no hope was held out for them when their badly battered boat was discovered on the rocks some three kilometres below Boar's Head Light. Arthur Covey and his fourteen-year-old son Charles of Indian Harbour, near Peggy's Cove, according to the *Halifax Morning Chronicle* of August 27, perished on their small schooner, the *Sligo*; the vessel ended up a broken wreck in Prospect harbour, the boy's body lying "prone and lifeless" in the cuddy.

The *Ottawa Citizen* of August 26, which described the gale as "one of the worst storms in the history of Nova Scotia," claimed that sixteen vessels had gone down in Louisbourg harbour, and in Halifax two schooners piled up on shore and a four-masted vessel, the *Veronica*, dragged her anchors and crashed into a nearby schooner, the *W. H. Eastwood*, "fouling her badly." A most unfortunate loss of a different kind was recorded in the *Halifax Herald* of August 27. The Newfoundland schooner *Vivian Ruth*, caught in the storm off Flint Island on the east coast of Cape Breton near Louisbourg, sprang a leak and began to sink. Captain Yarn and the crew of five were successfully transferred to a swordfishing vessel, but the battered hull took fire and the vessel foundered. The precious cargo of rum and whiskey was valued at fifty thousand dollars.

❦ ❦ ❦

By August 29 the vanguard of the banks fleet was arriving back in Lunenburg, with "tales of miraculous escapes, of heroic rescues, and of schooners making port with their crews lashed to the pumps and captains lashed to the wheel." The crippling of rigging and the loss of gear meant thousands of dollars in damage. What's more, the *Halifax Herald* noted, the fishing season would be shortened because of the gale. An item from Lunenburg on the same day was more sanguine. Seventy-five vessels of the fleet had not yet been heard from. Little anxiety was felt, however, since some of these must still be fishing off the Grand Banks and elsewhere. Even under ordinary conditions, these vessels would not be heard from for some time yet.

But damaged vessels were returning to other ports along the coast with similar stories of terror at sea. A man named Lovitt Surette, crewman aboard the Liverpool schooner *C. D. Zwicker* wrecked near Petite Riviere during the gale, heroically rescued his fellow seamen, Captain Dean Fralic and Ernest Roy, who could not swim, by carrying a line gripped tight between his teeth from the wrecked vessel to the shore. In Cape Breton a schooner named *John Halifax* was driven ashore at North Sydney. Built in 1862, this very same vessel had been cast up on nearby Kelly's Beach in "the infamous August gale of 1873," and was likely the sole survivor of that terrible storm. It was not greatly damaged this time around.

Reports were also already coming in from St. John's of the "calamity" that had occurred when the storm struck Newfoundland a day after it devastated Nova Scotia. There were staggering property losses and estimates that over fifty persons were believed to have "succumbed to the wrath of one of the worst storms to have ever visited this old colony." Ocean liners arriving in New York reported winds of 130 to 160 kilometres an hour which had churned the sea into waves twelve to fifteen metres high.

The *Halifax Herald*'s editorial of August 30 spoke of "Terrific Experiences": The Lunenburg banks fleet was known to have suffered severely in the recent storm, though "no loss of life has been reported at this writing....It is a hard calling, and hazardous, the calling of the Banks fisherman. After every storm Nova Scotia waits in dread for tidings that may come out of the wastes of the North Atlantic."

Bad tidings came all too soon. On September 6, almost two weeks after the storm, the *Bridgewater Bulletin* reported that two schooners of the Lunenburg fleet, the *Joyce M. Smith* and the *Clayton W. Walters*, were feared lost in the gale. They were known to be fishing in the vicinity of Sable Island at the time, and anxiety for their welfare was increasing. The *Joyce M. Smith* had been seen by the *Edith Newhall* fishing in ten fathoms off the Northeast Bar of Sable on the day of the storm. Captain Frank Risser of the *Marshal Frank* on returning to Riverport said that he had sighted a vessel's main boom off Sable Island but wasn't able to tow it. One of the crew had worked on the *Clayton W. Walters* and he thought the paint markings on the spar looked familiar. On the other hand, the schooner might simply have gone elsewhere to fish, as others in the fleet had done. Two men on the *Marshal Frank*, Bernard Mossman and Dawson Knock, had been knocked overboard and carried back again by the terrific seas, uninjured and none the worse for their experience.

Like the *Marshal Frank*, the *Partana* had been fishing southeast of Sable Island when the storm broke and put into Halifax reporting damage. Captain Guy Tanner, having come upon a main boom with one end submerged and apparently attached to wreckage, feared that these items belonged to some schooner that had gone down. With only 1,500 quintals of fish on board, the *Partana*, after repairs, returned to the banks to complete her trip.

The *Halifax Herald* of September 7 remarked that "the deepest anxiety prevails" respecting the safety of the two Lunenburg schooners and their crews of more than forty men: "They were staunch craft and in charge of captains who may be reckoned to give the utmost skill in times of peril." Both Captain Edward Maxner of the *Joyce M. Smith* and Captain Mars Selig of the *Clayton W. Walters* have "always borne reputations for skill and prudence, and if their craft did meet disaster, it may be taken that it was through no fault of their own." There was a ray of hope, the newspaper claimed, as the aptly named fisheries inspector, Ward Fisher, had received a message from the Canadian government steamer *Arras* on August 31 saying that a handliner, probably the *Clayton W. Walters*, had been seen on the St. Pierre Bank. But the *Marshal Frank*'s sighting of wreckage off Sable Island was more convincing. The newspaper had also been informed by a member of the firm W. C. Smith & Company in Lunenburg that flotsam had been picked up by the Imperial Oil Company steamer *Albertolite* off Cranberry Head near Canso, including a sea-chest containing personal effects and a letter supposedly addressed to one James Warner on the *Joyce M. Smith* (there was no Warner on board, but there was a James Warren). A waterlogged dory with oars inside had also been found. Frank Selig, brother of the captain of the *Clayton W. Walters*, noted in the *Halifax Morning Chronicle* of September 5 that the vessel had been out nearly two months and should have caught enough fish by now; he was anxious because she had not yet returned to port. Two days later the newspaper reported that the young wife of the captain entertained no hope that her husband or his vessel had escaped the storm. "If Mars were alive," she said, "he would contrive to let me know."

In its editorial on September 6, the *Bridgewater Bulletin* remarked that it was only twelve months since the loss of the *Sylvia Mosher* and the *Sadie A. Knickle*. Thoughts then had been directed towards safety devices. Member of Parliament W. G. Ernst had repeatedly urged in the House of Commons that the government equip one of its wireless stations for radio broadcasting to send weather reports every day at certain hours verbally instead of in code. Each vessel should have an ordinary radio receiving set, the cost of which would be nothing compared to the loss of life and property. The handline fleet that operated in the vicinity of Sable Island was in particular danger; if they had some warning, Ernst argued, they could obtain sea-room and not be at the mercy of the lee shore. On April 9 Minister of Marine and Fisheries Arthur Cardin had finally assured Ernst on the floor of the House that the service was being undertaken at Louisbourg. One might "reasonably expect it might be made good... but 'Nothing Has Been Done,'" exclaimed the *Bulletin*. "Human lives are too precious for trifling."

The vessels of the fleet were fast returning to their home port, but now three schooners appeared to be missing. There was a gleam of hope when four Lunenburg vessels were reported off Halifax, but they arrived in port without the missing vessels among them. On September 14 the *Lunenburg Progress-Enterprise* suggested that the third missing schooner, the *Mahala*, might be all right, for not so much was known about her movements as the others. However, she had been out a long time and fears were growing stronger that she'd foundered. In an editorial the newspaper remarked that the August storm had evidently taken its toll on the men in the Lunenburg fleet: "Over sixty men in peril, or lost at sea. The prayers of anxious hearts go out for their preservation and safe return, and it is fervently hoped that the hand of destiny has not for the second time in two years invaded the cottages where the wives and children of the fishermen are looking and anxiously waiting for the home coming of their loved ones and bread winners." Disaster had clearly befallen the fishing industry once again. "The gloom and anguish of loss has now settled over the little village of Blue Rocks, from which nearly the entire crew of the *Mahala* were shipped," wrote the *Bridgewater Bulletin* on September 13.

Earlier, on September 5, the *Halifax Herald* had mentioned that there had been some fears for the safety of the schooner *Haligonian*, Captain Moyle Crouse, but she had turned up; like the champion *Bluenose*, she was "too good a vessel for the gale to break." The same day the newspaper commented that there was also some anxiety about the other great racing schooner, the *Columbia* out of Gloucester, but Captain Ben Pine had sent a message stating that she was salt fishing and was not due home until about October 10. He believed she was safe.

The steamer *Arras*, Captain Clement Barkhouse, had been searching for the missing vessels since the "big blow." In mid-September, at the request of those involved in the industry in Lunenburg, the government sent the *Arras* out around the Western Bank and Sable Island to look for wreckage or for any disabled vessel. On board were six Lunenburg skippers: Captain Albert Selig, representing W. C. Smith & Company, Roland Knickle, representing the Lunenburg Outfitting Company, Eric Corkum for Zwicker & Company, Henry Winters, Acadian Supplies, Angus Walters for Robin, Jones & Whitman, and Captain Albert Knickle for Adams & Knickle. In due course Captain Winters sent a report to the *Morning Chronicle*: "Have thoroughly searched Sable Island for wreckage and found nothing that we could identify as belonging to any Lunenburg vessel." The fate of the schooners remained a mystery.

On September 21, almost a month after the storm, the *Morning Chronicle* informed its readers that in fact there had been no word from a fourth schooner, the *Uda R. Corkum*. Her owners, according to the *Halifax Herald* on September 13, had understood that she had left Burin in Newfoundland for the Grand Banks about three and a half weeks earlier, but Captain Freeman Corkum, one of her owners and previously captain of the vessel, had now been informed that she had been fishing near Sable Island at the time of the storm. She should have reported home days ago. "[T]here is slowly dawning on the people of this town and community that the 'Grand Fleet' so long the pride and the very mainstay of Lunenburg has suffered the worst disaster in its history," lamented the *Chronicle*. The men had been "taken in the very noon of life." Eighty men lost. It was as great a loss in one season's fishing, the newspaper observed, as the town had suffered in the whole four years of war.

The wreckage of the *Uda R. Corkum* was ultimately located by Captain Barkhouse of the *Arras* on September 27 on the northern part of the Middle Ground in a place known by Lunenburg fishermen as the "Sixteen Fathom Spot," about seventy kilometres off Sable Island. Captain Scott Corkum of Acadian Supplies, managing owner of the vessel, said the schooner had last been reported on the western shoal of Quero Bank; he didn't think the vessel had run afoul of Sable Island and speculated that "there might be something in the theory that when a storm hurls the water in huge waves, piling them up for many feet, and a vessel sinks into the trough of the sea, that if she happens to be in shallow water she may strike bottom and thus come to her grief." Whatever had happened, the *Arras* was able to salvage the wreck and bring back to Lunenburg the main gaff, topmast, and main boom. The vessel had been built for Captain Freeman Corkum and he had sailed on her for so many years that he was familiar with every bit of tackle on her; he was therefore able to identify the wreckage. Thus, declared the *Progress-Enterprise* on September 28, the vessel was identified beyond doubt.

It had been a month of harrowing uncertainty, waiting to learn the fate of the vessels in the fleet.

On September 24, 1927, at the time of the annual festival of "the Harvest of the Sea in Lunenburg, the Home of the Deep-Sea Fleet" (in short, the fishermen's picnic), the *Halifax Herald* paid a special tribute to the lost fishermen. Beneath a boxed-in headline and a list of the four Lunenburg vessels lost and their captains, the item referred to the families "who will watch in vain for the sight of familiar topsails appearing over the headlands."

Ships That Never Come In

Never again will these gallant vessels ride at anchor in the home haven.

The familiar surroundings that knew their devoted masters and crews will know those familiar figures no more.

Very few of us, one fears, realise what the life of our deep-sea fishermen means. It is so easy to sit in comfort and dream of the Romance of the Sea…so thrilling to picture white sails and tall spars, hardy men tanned by sun and spray, "clinging to reeling decks in the teeth of the living gale." Mere words—fancy. How often does the realisation go beyond mere words and fancy? Argosies are in story-books: The life of the fishermen is condensed into four lines:

"We bring no store of ingots,
No spice or precious stones—
 But what we have we gathered
With sweat and aching bones…"

It is a hard life and perilous—and harder and more perilous in these years of treacherous weather, when the glass "turns upside-down" while men watch it.

Too often, one thinks, we reserve our admiration for the heroes of romance, of derring-do, forgetting the courage and resourcefulness of men whose lines are cast in the "commonplace" pursuits of commerce, who take their lives in their hands every time their vessels clear for the fishing grounds.

Not every man makes a good or successful fisherman. It is an occupation that calls for a high degree of skill and intrepidity. Sometimes the return is good—sometimes meagre. But never is the reward beyond what our fishermen deserve.

Of what does courage consist, if not of this determined devotion to duty in the teeth of whelming odds? For with Tragedy again in their midst, these capable men, undaunted by tempest and peril, will be preparing now to fare forth another season, to meet the worst the storm can bring – to give their lives, if need be, to the perpetuation of the industry that means so much to Nova Scotia.

These are the heroes—these are the bread-winners.

"Over Eighty Lives Lost in Great Disaster," pronounced the *Herald* headline in large bold type a week later, on October 1, devoting an entire page and more to the calamity. "Lunenburg Fishing Fleet Suffers Most Appalling Tragedy." The article, written by Agnes G. McGuire, a native of Lunenburg County, spoke of "unutterable pathos," of schooners "blotted from the face of the waters with all on board." There were photographs of the captains who'd been lost, and photographs too of "stricken homes" in the village of Blue Rocks. The front page was adorned as well by a lengthy poem by J. H. McCulloch called "Sable Island":

I lurk out here, where bankers trawl:
An evil place, where no ship dare call.

No living thing thrives on my destitute sands,
And I laugh at the efforts of puny hands
   To warn men away from me.
They have reared their light on my shifting heights,
But the night always comes when the doomed ship fights
   Through the gale on my storm-hidden lee.
...

In my quick yellow sands they disappear,
The white-winged ships that the gales drive here:
Tall China clippers and French Brigantines,
Lunenburg bankers and Cape Cod lateens.
   And their crews, so valiant and fine
Five hundred tall ships I have clutched and buried,
A thousand men to their long sleep I've hurried,
   In these sinister sands of mine.

Agnes McGuire, in full flight, called Sable Island "that loathed and feared spot...the greatest of all menaces to the fishing fleet." "This grim land," she wrote, "although plausible enough in its attitude toward the sailors in fair weather, is almost credited with fiendish qualities when its mood changes and it crouches, octopus-like, with its tentacles thrust under the sea for miles and its 'live' sands, as the sailors call them, shifting for miles so that only the log can testify to the duplicity that is being carried on below the surface....Do the sailors hate the echo of its wrathful roar and the melancholy mockery of its booming surf? There is not one of them that nears its shoals, with its plentiful harvest of fish on those feeding grounds, but knows that he is venturing into the very Vestibule of Death."

Dead men cannot tell their tales, but the men who survived the ordeal can give us some idea of the horrors of the storm. Captain Leo P. Corkum, skipper of the vessel *Maxwell F. Corkum*, recalled how quickly the storm approached: "[W]e were fishing off Sable Island when the glass suddenly began to drop. I must say in all my thirty-nine years as a skipper, I never saw anything like it. I blew the distress signal and the dories hauled their trawl as quickly as possible....It was a beautiful morning with just a nice breeze...." Only those in a similar situation, he maintained, would know the "awful fear and the terrible pounding." After the gale, not a dory or shred of sail remained, and the booms were as clean as if no sail had ever been there. "White sand filled every crevasse." Captain Corkum said the deck had been swept as clean as if it had been "holystoned by a navy crew." The men brought out extra sails and they made their way into Canso.[2]

The *Edith Newhall*, Captain Gordon Mosher, also narrowly escaped disaster. Fishing near the Northwest Bar of Sable Island when the gale struck, in order to save his vessel and the crew the captain was forced to choose a course directly over the bar and in eleven fathoms of water. The perilous passage was made safely, but the schooner was badly damaged, according to the *Halifax Herald* of August 31. The enormous seas that swept over the vessel broke the main gaff and smashed the hatches and skylights in the cabin and engine room; part of the vessel's rail was smashed, stanchions were knocked loose, and the light box and lights in the rigging were carried away. The vessel was leaking badly when she finally arrived back in port.

Another Lunenburg vessel that crossed the bar that August night in 1927 was the *Andrava*. Early in the evening the mate Lemuel Isnor, from Indian Point near Mahone Bay, roused the captain, Roland Knickle, from his bunk with the ominous words: "The sky looks peculiar, Captain. I think we're going to have a breeze." The schooner had left Canso in clear weather the day before with a crew of eighteen on a routine fishing trip on the Western Bank, and she found herself about six and a half kilometres south of the West Light on Sable Island when the storm struck. With the gale steadily worsening, the vessel too close in shore, the captain decided to gamble; he changed course and sailed west towards the island, hoping to reach the calmer and deeper water on the north side by taking the vessel across the submerged bar. The raging seas destroyed sails, broke spars, and swept dories off the deck. One crewman, Austin Knickle, pinned under a fallen chain locker when the vessel pitched, cried out for help but the roar of the storm muffled his calls and he was only discovered and released hours later. Enos Burgoyne was up forward when a large sea boarded the vessel, throwing him back to the cabin house; he managed to catch hold of a ring bolt, thus saving himself from being washed overboard. A second large wave later in the evening heeled the vessel far over but she righted herself. Isnor volunteered to be lashed to the wheel. Leaking badly, the vessel crossed the bar with great lurches, each huge wave in turn lifting the schooner then plunging her down till the keel grazed the sandy bottom. By three in the morning they could hear the seas breaking behind them. They had made it. The deck, they discovered to their astonishment, was covered with sand. When the wind moderated, they started home for Lunenburg. Captain Knickle, with his vast seafaring experience—winter fishing out of Gloucester, sailing across the ocean and to southern ports—claimed that this gale was the worst he had ever encountered.[3] Thereafter, according to his great-niece Nancy Selig, he carried with him on his voyages a bottle of sand that he'd gathered up from the deck of the *Andrava* on that fateful night.

One man proved to be very lucky. Rector Mason of Lunenburg, according to his granddaughter Paula Masson (who spells her surname like the original settlers), had been fishing for some time on the *Mahala* but he left in June 1927 because the men had "bad mouths"; a God-fearing man, he didn't care for the men's swearing. Instead, Mason sailed on the *Andrava*, and thus, fortuitously, instead of meeting his maker on the *Mahala* he survived the 1927 gale by crossing the Sable Island bar on the *Andrava*. Rector Mason dory-trawled for forty-seven years.

Captain Angus Walters was fishing in the *Bluenose* on the Misaine Bank at the time of the storm. "There was no canvas ever made strong enough to stand such a gale," he maintained. The *Bluenose* too arrived home damaged: she had lost her anchor, part of her hawser, her sails were ripped, and loose gear about the decks had been swept overboard. At times, according to a report in the *St. John's Daily News* of September 6, the sheer poles had been laying flat on the water under the force of the gale. "In all my seagoing experience," said Walters, "I have never seen the barometer go down and come up as quickly as it did on that occasion."

Lines of poetry in Harry Hewitt's history of Lunenburg describe the relief of safe arrival home:

> Safe home, safe home in port!
> Rent cordage, shattered deck,
> Torn sails, provisions short,
> And only not a wreck:
> But O the joy upon the shore
> To tell our voyage-perils o'er!

One of the first vessels to be confirmed lost in the August gale of 1927 was the *Joyce M. Smith*. Built by J. N. Rafuse & Sons, at Salmon River, Digby County, in 1920 for W. C. Smith & Company, she was registered in Lunenburg at 102.5 tonnes net register and a length of 37.3 metres.[4] The captain, Edward Henry Maxner, fifty-five, of Lunenburg, was also her managing owner and he had sailed her from the time she was built. There were three Lunenburgers aboard: the captain and his son William, twenty-four, and John Peterson, age sixty-eight, the cook (some sources say he was the salter). The rest of the crew came from Newfoundland and their fate will be

Captain Mars Selig, Vogler's Cove, and his wife Hazel.

explored in the following chapter. Captain Maxner was married to Floresta Corkum, who died in 1962, thirty-five years after her husband met his death on the *Joyce M. Smith*.

A second shipwrecked schooner, the *Clayton W. Walters*, had been built as a banks handliner for Captain Stannage Walters of East LaHave in 1916 by the McGill shipyard in Shelburne. Eventually she was purchased by Mars Selig, who had sailed as mate with Captain Walters. The vessel was 28.3 metres long and 72.5 tonnes and she was provisioned by the Lunenburg Outfitting Company.

Captain Marsden Selig, age thirty-three, from Vogler's Cove, a village down the shore and on the other side of the LaHave River from Lunenburg, was in his second year as master aboard the *Clayton W. Walters*. His wife, Hazel, was pregnant when he left for the banks and their son Paul was born on July 16, 1927. His father would never see him. Hazel raised her two children alone, never remarried, and as an elderly woman spoke to Ralph Getson of the fisheries museum in Lunenburg before she died, aged 102. Mars Selig was one of fifteen children. His father, Captain Adam Selig, had died tragically just seven months before the gale struck; helping Mars unload some firewood, he fell from the top of the load, landing on his head on the frozen ground. Also on board the *Clayton Walters* were Mars Selig's first cousin Raymond and a young nephew named Guy. Both men, like the captain, were from Vogler's Cove. Guy Selig had replaced his father, who was ill, and he planned to make this his last fishing trip, hoping to save enough from fishing and picking apples to attend barber school in Halifax.

Guy Selig, Vogler's Cove, Lunenburg County.

Alfred Conrad, Mars Selig's brother-in-law, also from Vogler's Cove, was survived by his wife and six children. His brother Selborne, who lived in Boston at the time, had planned to go with Alfred on this fishing trip, but he was offered work in the city and remained there; Selborne would live to be ninety-seven, dying in Caledonia, Queens County, in 1994. The *Halifax Herald*, on October 1, 1927, noted that Vogler's Cove was suffering once again "the mute testimony of common grief and the community stands aghast at the number of

Leslie Oickle (left), Lower Sandy Point, Shelburne County, and a buddy.

Raymond Williams (left front), West Green Harbour, Shelburne County, and friends; Foster McKay (middle back row) died on the *Columbia*.

children deprived of their fathers and the doubly and trebly bereft families caused by brothers and relatives sailing together."

> ### LOST ON THE *CLAYTON W. WALTERS*, AUGUST 24, 1927
>
> **FROM LUNENBURG COUNTY:**
> **Captain Marsden Selig**, 33, Vogler's Cove
> **Guy Selig**, 21, Vogler's Cove, nephew of the captain,
> **Raymond Selig**, Vogler's Cove, first cousin of the captain
> **Alfred Conrad**, Vogler's Cove, brother-in-law of the captain
> **Victor Anthony**, Vogler's Cove, cook
> **William Robert Himmelman**, 28, Mount Pleasant
> **Percy S. Himmelman**, 35, Mount Pleasant, brother of William Himmelman
> **Roy St. Clair Hiltz**, 26, Mount Pleasant, brother-in-law of William and Percy Himmelman
>
> **FROM QUEENS COUNTY:**
> **Otto Reinhardt**, East Port Medway
> **George M. Smith**, East Port Medway, father-in-law of Victor Anthony
> **Reuben M. Smith**, East Port Medway, brother of George Smith
> **Fred Whynot**, East Port Medway
>
> **FROM SHELBURNE COUNTY:**
> **Warren Conrad**, 17, Jordan Falls
> **Morton Enslow**, West Green Harbour
> **Archibald Enslow**, West Green Harbour[5]
> **Atwood Firth**, 33, Jordan Falls
> **Jacob Hartley**, 27, Jordan Falls
> **Leslie Oickle**, 18, Lower Sandy Point
> **Bradford Williams**, Jordan Falls, brother-in-law of Atwood Firth
> **Burns Williams**, 63, West Green Harbour
> **Raymond Williams**, 21, West Green Harbour
> **Gordon Williams**, 17, West Green Harbour, brother of Raymond Williams

Two young brothers from West Green Harbour in Shelburne County, Gordon and Raymond Williams, died on the *Clayton W. Walters*; it was Gordon's first trip to sea. Jacob Hartley, from Jordan Falls, well known as a hunter and trapper, also perished in the gale. Leslie Oickle, from Lower Sandy Point, had been in the Royal Naval Canadian Volunteer Reserve; his younger brother Amos had planned to go with him on the *Clayton Walters* but at the last minute, at the railway station on the way to the vessel, he had decided not to go.

Another Shelburne County man lost in the same storm, Manus Hemeon from Lower Sandy Point, was washed overboard from the schooner *Julie Opp II* out of Lockeport; crewmate Horatio Enslow, reported in the *Shelburne Gazette and Coast Guard* of October 20, said that Hemeon had been lost when he went up to secure a barrel that was rolling about on the deck.

Captain Warren Knickle, Lunenburg.

"The picturesque village of Blue Rocks, with the laughing waves dimpling in the sun which bathed the grim rocks and shores," wrote Agnes McGuire in the *Halifax Herald* on October 1, "at first gave no hint of the intense travail of mind within the cottages." But now "an air of poignant grief" overshadowed and pervaded the settlement and the stricken families, for it was becoming known that the *Mahala* had foundered, with almost twenty men either from Blue Rocks or with strong Blue Rocks connections. The *Bridgewater Bulletin* remarked on September 27 that the loss of nearly twenty of "its most stalwart young men...means taking away almost the entire family support." A reporter for the *Halifax Herald*, on October 10, wondered if Blue Rocks "perhaps was the hardest hit of any place...Seldom does a place of this size have to mourn at one time such a loss of manhood, and never in the history of the industry has such an appalling loss of life taken place as this year....Flags were dropped at half-mast all along the Lunenburg waterfront and from the masts of the vessels in mute tribute to the bereaved households."

One of the newest vessels in the fleet, the *Mahala* was launched at Mahone Bay at the John McLean & Sons shipyard late in 1925 for Captain Warren Knickle. This was her second year fishing. Outfitted by Adams & Knickle, she was 37 metres long and 88.9 tonnes. Henry W. Adams was her managing owner.

A young Ronald Knickle, Blue Rocks, Lunenburg County, with his family.

The family connections of the men on board the *Mahala* were intricate, to say the least. Captain Warren Knickle, son of Archibald Knickle of Lunenburg, had first sailed as master at age twenty-four and would be lost at sea on the *Mahala* only four years later. His two brothers, Granville and Owen, were also on board the *Mahala*, as cook and second hand respectively. Also on board was the captain's brother-in-law, Scott Miller, whose wife therefore lost her husband and three brothers in the tragedy. Granville Knickle's wife lost her husband, his two brothers, and three brothers of her own, Irving, Wilfred, and George Tanner. William Tanner and Archibald Knickle each lost three sons. Russell Mason, who went down on the *Mahala*, was the son of Blue Rocks merchant George Mason. Ronald Knickle was only sixteen. Lee Knickle had been married only three weeks.

Granville Knickle's daughter, Grace Veinotte, interviewed eighty years later in 2007, said that her father was on the *Mahala* as cook while waiting for his own vessel to be built. "Life," she declared, "never returned to normal.... They never got over it. My grandmother sat by the kitchen window until the day she died, looking for the boat to come in."[6]

With three *a*'s in her name, the *Mahala* was supposed to be a lucky vessel. The three strokes in the capital A represented the Trinity, which was said to bring good fortune, and three *a*'s were better than one (as were three *e*'s). But unlike the *Andrava* and the *Marshal Frank*, or the *Theresa E. Connor*, the *Mahala* was not lucky. At her launch, one of the workmen broke his nose and shed blood; said Ralph Getson, this was taken to be a bad omen.

Lee Knickle, Blue Rocks, Lunenburg County, and his mother.

Blue Rocks and nearby Stonehurst were in fact generally short on luck. Three young men from the area, Aubrey J. Knickle, Wesley Whynacht, and Heber Miller, had been lost early in 1924 on the *Keno* with Captain Albert Himmelman on a voyage to Newfoundland for frozen bait. Five more local men had gone down on the *Sylvia Mosher* in August 1926. That same month two men, Forden Weaver and Currie Greek, had drowned off Blue Rocks when they attempted to swim to the nearby shore from a capsized dinghy; local boy Ellsworth Greek saw the men struggling in the water and went to their rescue in his dory, managing at least to save Blake Heinick (Hynick). It was a "pleasure cruise in the harbour" that went wrong, said an item in the *Lunenburg Argus* of August 19, 1926. At least one man from Stonehurst died on the *Uda R. Corkum* in August 1927, on the same day the men from Blue Rocks went down on the *Mahala*.[7]

Lost on the *Mahala*, August 24, 1927

FROM LUNENBURG:
Captain Warren Knickle, 28
(James) Granville (Archibald) Knickle, 33, brother of the captain, cook
(John) Owen Knickle, 26, brother of the captain, second hand
Scott Miller, 39, brother-in-law of the captain
Owen Risser, 25, originally from Blue Rocks

FROM BLUE ROCKS:
(Albert) Irving Tanner, 34, brother of Granville Knickle's wife
Wilfred L. Tanner, 28, brother of Irving Tanner
(George) Sammy Tanner, brother of Irving Tanner
Reginald Hynick, 50

Charles Hynick, 19, son of Reginald Hynick
Mack (Maxwell St. Clair) Knickle, 23, cousin of the captain
Ronald S. Knickle, 16, throater, nephew of Maxwell Knickle
(Norman) Basil Knickle, 26, cousin of Ronald Knickle
Fraser Knickle, 19, header, brother of Basil Knickle
Lee Knickle, 23
Russell Mason, 28
Milward A. Greek, 30
Fred Romkey, 25
Stafford Urias Romkey, 33, related by marriage to the captain

FROM FIRST SOUTH:
Albert Lohnes, 33

Captain Wilfred Andrews, Indian Point, Lunenburg County.

The *Uda R. Corkum*, the last of the Lunenburg vessels to be identified as missing, was built in 1918 for Captain Freeman Corkum at the Smith & Rhuland shipyard in Lunenburg. Captain Wilfred Andrews, age thirty, was a veteran of the First World War, having served overseas for three years. He had been married to Elva Louise (Langille) only two years before his death. The *Uda R. Corkum* was 34.3 metres long and 90.7

tonnes. To compare, the better known *Bluenose* was registered at 39.6 metres and 89.8 tonnes; the *Theresa E. Connor*, the last of the bankers, preserved at the fisheries museum in Lunenburg, is 42 metres long and 82.5 tonnes.

Almost all of the men on the *Uda R. Corkum* came from the Mahone Bay area, especially from Indian Point. As the *Halifax Herald* noted on October 1, one mother in Indian Point was mourning the loss of "three splendid sons," and it was merest coincidence that her two other sons did not sail on the *Corkum*. The five Andrews boys, sons of a widowed mother, had bought a number of shares in the schooner. Locally they were known as "striving" men. Wilfred, who was to go on the trip as master, proposed that they all sail together, thus benefiting each other if all went well. But one of the older sons, James, demurred, saying it was not right for all to go on the same vessel. So Wilfred went, along with Samuel, who was married and living in the old homestead, and Reginald, age twenty-one, who was always

Oran Eisnor, Indian Point, Lunenburg County.

Reginald Andrews, Indian Point, Lunenburg County.

referred to by his mother as "My Baby." The two remaining brothers sought other berths. At the time of the tragedy, James Andrews was on board a schooner some 480 kilometres out from Burin, Newfoundland, and did not even know there'd been a hurricane. Only when he arrived at Canso did he have any fears for his brothers.

Also from Indian Point was Leslie Heisler, who left behind two orphans, their mother having died about a year previously. George Hiltz, from The Narrows at Indian Point, was survived by his wife and six children. It was the first trip to sea for George and his son Reuben, as they'd been unable to find work on land. Mrs. Hiltz was apparently ill for two years after the loss of her husband and son.

LOST ON THE *UDA R. CORKUM*, AUGUST 24, 1927

FROM INDIAN POINT, LUNENBURG COUNTY:
Captain Wilfred A. Andrews, 30
Samuel Andrews, 32, brother of the captain
Reginald Andrews, 21, brother of the captain
Oran Eisnor, 22
John Eisnor
Leslie Heisler, 40
Perley M. Heisler, brother of Leslie Heisler
Wilbert K. Wentzell, 23
George W. Hiltz, 54
Reuben C. Hiltz, 17, throater, son of George Hiltz

FROM ELSEWHERE IN LUNENBURG COUNTY:
Clarence Wentzell, 22, Clearland
Davis A. Mills, 17, Oakland
Norman Eisenhauer (Eisnor), 55, Mader's Cove
Henry Hamm, 52, Mader's Cove
Elvin Tanner, 24, Stonehurst
(William) Isaac Jennings, 47, First South, cook

FROM NEWFOUNDLAND:
Thomas Nolan
William Strawbridge
George Gilbert

Two men from Stonehurst, William Tanner and Fred Wagner, may have been aboard, but this has not been confirmed.

A service in memory of the twelve men of the parish who died on the *Uda R. Corkum* was held on a Sunday in October at St. John's Evangelical Lutheran Church, Mahone Bay. The *Bridgewater Bulletin* of October 18 reported that the Rev. E. V. Nonamaker was assisted by Rev. Cater Windsor of the United Church and Rev. E. Paul of the Baptist church.

"Not all the sons of Lunenburg who have run their course on earth lie beneath the green sod of the old cemetery in the hometown," mused Harry Hewitt in his history of the town. "Many there be who sleep their last sleep full many a fathom deep. Not all the vessels built at Lunenburg reach an honored old age, and moulder away in the coves and recesses of the harbor and surrounding inlets. Many there be of these skeleton frames of which provide hiding places for the denizens of the deep. Vessels and men alike have often fallen prey to the angry elements, despite a valiant conflict by both."

We have too few opportunities to hear the voices of the fishermen who went to sea on the schooners or the words of their families back home. An exchange of letters among the members of the Tanner family in Stonehurst, the small community by the sea near Lunenburg, is therefore of great value. Their love for one another and the longing to be together are palpable.

The correspondence begins on September 10, 1913, with Willietta Douzella Tanner writing to her husband, Daniel Austen Tanner, from her home in Stonehurst.

> Dear husband, I now set down with love and pleasure…
> it is over a month since i herd from you. i herd that you was not feeling very good…you are gone over three months. the children is anxious to see you…A worry…all summer just thinking about you. we are all smart at the present time. Dear Husband i hope that you will soon come home to your dear loveing ones.

On September 18 Willietta heard from her husband:

i must till you that we had fine weather on the banks. but we did not do much we only got six hundred. i am in a herrey and you must excused this letter and riten my Dear…i must come to a close by wishing you all a loving good night so no more from your sweet heart. Mr. Danieal Tanner to his best love Mrs. Danieal Tanner. i kud rite a little more but I ant got the time please ancer to Cape Broyle New fond land Sch. Uda A. Saunders.

Willietta responded on October 8:

I only wish for the time to go around quick for it seems so lonely home here.

By 1917, four years later, Daniel and Willietta's son Elvin Alfred Tanner, age fourteen, had joined his father aboard the schooner *Elsie M. Hart* as flunkey. Willietta sent a rather plaintive letter from Stonehurst on July 2.

I now sit down and write you a few lines to let you know we are all smart hopeing that you and Elvin are the same for I don't know if you smart or if you are sick for I diden't get any letter from you yet all the rest of the women got mail saturday but I did not get any…well yesterday was Sunday and it was so nice I wish you could of been home with us to spend the Sunday I knew it would of been more than pleasant for us for it is so lonely home here with all of you gone that we will be more than glad when you can come home to us again which I only hope you will all come back safe to us and with a good trip of fish.

By July 18 Willietta had heard from her husband and son.

I only hope it will get nice so if you go out you will not have to fish in…ugly weather for I dread it so when it is ugly & blowing. I am all the time thinking about you…well I hope Elvin is a good boy and listens to you and uses the cook good.

On July 19 Elvin's sister Olive wrote to him:

guess you see lots of nice places down round there. I wish I could see some of the places and I guess you have a good time too when you are in a harbour with the girls ha ha. well I hope you are a good boy and listens to papa and you must be good to the cook.

Olive added that she hoped he was getting lots of "tongues and sounds" because she was hoping for enough money to buy a new dress next fall.

By May 22, 1918, Elvin, now fifteen, was a throater aboard the *William C. Smith*, and wrote to his mother from "Madglean Island, Low Point." He had had a sore thumb and "diden bouther to right," and they'd had a feed of lobsters at ten cents apiece. He liked being a throater and the captain was very good to him. He told his mother to tell Victor, his younger brother,

that papa said he should be good to mama and Olie and not to go in the boats Because I dreamt that he was drouned I thould I could see him lay in the sea grass. I hope it aint true.

By May 12, 1920, young Victor had become flunkey aboard the *William C. Smith*, a third family member for Willietta to worry about.

Listen at papa and be good boys. I hope it won't be so stormy on the banks so papa dosen't have to work so hard...We find it so lonesome Olive & I We thought when night come Victor had to come so use to him being home with us. And we miss you and papa too so much...we feel so lonesome for you all I wish you could be all home to night with us for it is so nice home here to night.

On August 14, 1921, Elvin wrote to his mother from St. John's.

I now drop you a few lines to let you know that we arrived in St. Johns N.fld. with twenty two hundred Centls and all hand well and good health thank God there are some rought weather and fine weather. well we will go out this time and that will end our trip we are taking 2 thousand of Squid and 60 hogs of salt. there are some fish on the Banks and Squid. if we had all fine weather to Catch them...Well I haven't much news to tell you this time because it is the one thing over. and over. But long as we are smart and heaty we are all right.

The final letter in the collection is from Elvin, on board the schooner *Jean J. Smith*, sent from Cape Broyle on July 4, 1922:

Dear Mother...all of us are smart hoping you a Olive is the same. well the weather is very ugly not so much wind but a lot of fog fish is very scarce. well we are baited and will be leaving for the banks...the news is not very plenty when there is no fish ther is nothing to write but I thought I would have to write anyway...dont forget to put 4 cents stamps on your letters so they will reach Newfoundland...We diden see any sun for two weeks and I guess we wont see any till we get home. there were some squid this time on the bank and I guess we won't get in St. John this summer. Well I guess I must soon close. So good Night till we see you again. From your sons and loving husband. love to all.⁸

Elvin Tanner perished on the *Uda R. Corkum* on August 24, 1927, in his twenty-fourth year. He left behind his wife, Helen, and a baby boy, born only six days before Elvin's death at sea. Willietta Tanner died the following year at the age of forty.

On October 26, 1927, the *Halifax Herald* reported that the prices for fish were better than in the previous year, but only slightly. In 1926 the price for the first two trips had been $5.50 to $6.00 and $5.50 for the summer trip. This year the frozen bait trip brought $6.35 per quintal and the spring trip $5.80 to $6.40. It was expected that the summer trip would bring $7.00 a quintal. However, it was estimated that the cash value of the catch this year would be $300,000 less than in 1926. The fleet was also decreasing in size. There were eighty-three vessels in the fleet in 1927, including the four vessels that were lost, which was nine less than in 1926. This year the fleet had brought in 227,590 quintals of fish, over 115,000 fewer than the previous year. The loss of vessels and property had been appalling. The cost of schooners and their equipment and running expenses were very high. With profits small, people were not investing, and the fleet was not expanding as it should. Two new vessels were being built, but four had been lost. With no new contracts for vessels, shipbuilding in the town was at a standstill, at least for the time being.

Most serious, however, was the loss of life. The *Canadian Fisherman* commented in October 1927 on the terrible loss of six vessels in two years: "[W]hen the toll of precious lives is so appalling...it is then that brave hearts will consider the danger of their calling and many give up going fishing and seek less hazardous tasks. In this way will the industry be ruined for when the men, the backbone of the fleet, refuse to go, then the wooden ships will have to remain idle, and this is just what those interested in the fishing industry are beginning to fear." The loss of men and captains to man the fleet, together with the low fish prices, declared the *Halifax Herald*, was having a most discouraging and depressing effect. The future of the fishing industry was not bright.

At a meeting of the American fishermen's race committee on November 21, 1927, reported in the *Progress-Enterprise* of December 7, a resolution was adopted declaring that "the citizens of Gloucester have learned with profound sorrow of the foundering of four staunch Lunenburg fishing schooners with their crews of eighty-five men." It was an "irreparable loss," brought home by Gloucester's own losses, the schooner *Columbia* and her crew as well as the *Avalon* that had been lost in late October. In this hour of mutual sorrow, the committee members tendered to the people of Nova Scotia their "heartfelt sympathy and condolences," more accentuated by the fact that many Nova Scotia men had come to their city in the past half century and contributed much to its social and commercial structure and prosperity. "We knew them in life. We mourn them in their passing....So in this hour, we bow our heads to the decree of the All Infinite and commend the bereaved to the Divine consolation which has been our strength in ages past, echoing the simple prayer of the Breton fishermen.

> Lord the sea is mighty
> And our boats are small."

5
NEWFOUNDLAND
and the AUGUST GALE of 1927

THE BRUTAL STORM THAT RAVAGED NOVA SCOTIA ON AUGUST 24, 1927, and sank four Lunenburg schooners off Sable Island moved inexorably on out over the Atlantic, leaving more destruction in its path. The *Halifax Herald* of August 27 reported that the shores of Newfoundland were strewn with wreckage, shipping had been badly hit by the storm, the fishing fleet at Bonavista had been destroyed, the schooner *Henry Dowden* was a total wreck near Trepassey, and a Burin schooner, identified only as the "McLaughlin," had been found bottom up in Placentia Bay with a man's body tied to the rigging. Three fatalities had been reported.

There was worse news to come.

In his introduction to *Outrageous Seas: Shipwrecks and Survival in the Waters of Newfoundland*, Rainer Baehre estimated the overall number of shipwrecks in Newfoundland waters from the sixteenth century to the end of the twentieth century to be in excess of ten thousand. These figures would include losses in many fierce storms before the 1920s: the thousands of mariners who died in the Great Newfoundland Hurricane of 1775; the Great Labrador Gale of

October 1885 in which seventy-five fishing vessels were wrecked and at least seventy lives lost; a fierce August gale on the Grand Banks in 1887. Newfoundland marine historian Robert C. Parsons in *Survivors and Lost Heroes* wrote that the August gale of 1892 was one of the worst: it was an intense storm that roared across the mouth of Placentia Bay and around Cape St. Mary's at the southern tip of the Avalon Peninsula, an area frequented by fishing vessels and coastal traders. He remarked that we now rarely hear about August gales, but before improvements in weather forecasting and long-distance communications "the sudden, intense fall wind storms were very much in the news," and much discussed by the men at sea and the owners of ships. These storms were usually the tail ends of tropical hurricanes, affecting a confined, localized area, and they "spread death, hardship and economic setbacks to mariners, their families and communities."

It was just such a storm that led to the death of many Newfoundland fishermen in August 1927.

The weather report in the *St. John's Evening Telegram* of August 24 read: "Toronto (Midnight)—Fresh south and west winds, cloudy, with showers." The next day: "Line trouble—no weather report received." There was, then, as usual, little or no warning. Indeed, the St. John's newspapers were initially much more concerned about events on the mainland, the *Telegram* fretting on August 25 that as yet there had been no solid news of the hurricane in Nova Scotia. The "fifty mile an hour gale," which had sprung up in the Maritime provinces the previous night and dislocated the whole of the telegraph service, reached St. John's in the early part of the morning, blowing from the west with considerable violence. "Despite the velocity of the wind, intense heat has prevailed and the cloudless sky all the morning makes it almost impossible to believe that a storm is raging." Some ships in the harbour had been dragged from their moorings, and those arriving had to anchor in the stream. On land some fences, trees, and windows had suffered damage. Rural telephone lines were out of commission, some poles down, and several farmers on the Mount Pearl Road had lost their stacked hay. The newspaper recalled the storm that had swept Cape Breton on the same date in 1873.

A day later, on August 26, the *St. John's Daily News* announced that there had been eight deaths in Nova Scotia during the storm, one of the worst in its history. Damage was believed to be heavy, but with telegraph and telephone lines down it was hard to get news from outside of Halifax. The tropical hurricane had been followed by a day of perfect summer weather. As for the "heavy gale" that had struck St. John's itself on the previous day, there had been considerable minor damage throughout the city and suburbs. In the harbour a big sea had raged, so several steamers had been unable to berth.

More distressing news was starting to come in from around the island, however. That same day the *Telegram* reported that the storm had caused havoc around the coast. The Belleoram schooner *Noxall*, Captain William A. Kearley, had foundered off Cape Spear with a load of coal bound from Sydney for Bay de Verde. Unfortunately Albert Cluett, who was at the wheel, had been washed overboard and drowned. The remaining crew had "the most trying time" keeping their dory afloat, and in the terrific gale they'd been driven some thirty kilometres out to sea. Nearly exhausted when they finally reached Flat Rock near Ferryland in the morning, they were put up at the YMCA to recover.

The *Telegram* also informed its readers the day after the storm that the schooner *Henry Dowden*, on her way from St. John's to Marystown, had run ashore at Biscay Bay near Trepassey. The crew was safe, part of the cargo saved, but the schooner was a total loss. Some twenty-five fishing boats had been smashed at Bonavista, putting about a hundred families out of a means of livelihood.

By the next day, August 27, the scale of the disaster was becoming apparent. "Thursday's Storm Fell Severely on Newfoundland's

Fishers," stated the *Daily News* headline. "Tale of Storm's Havoc Has Dire Proportions." Ten Newfoundland fishermen were known to have drowned, and there were fears for others. There were reports of loss of life and vessels and damage to property from all over the country, but the southwest coast had apparently been hit hard, with the largest number of deaths reported so far from Channel–Port aux Basques. The *Vienna*, John Chaulk master and owner, was lost with a crew of six, on a fishing voyage en route from the French island of St. Pierre to their home in Burnt Islands. Rails, bulwarks, and hatches, the woodwork of the engine room, and pieces of dories had been found off Fox Roost, along with two trawl kegs.

In an editorial on August 27 the *Telegram* described the disastrous toll of the storm, "which quite unexpectedly and with overwhelming force swept down from Nova Scotia." Second only to the loss of life, which the paper now estimated to be between thirty and forty "hardy fisherfolk," was the serious setback to the fishery, given the destruction of vessels and gear. "We feel certain that with that dogged resolution characteristic of Newfoundland fishermen they will not permit disasters even of such a serious nature as this to defeat them." The *Daily News*, in its editorial, urged its readers to think of those left behind, and strongly supported a relief fund: "Again the unsparing hand of death has brought tears and sorrow to the South West Coast, and six of Burnt Island's [sic] sons have found a watery grave. Yearly the tale is told, and sometimes the blow falls when least expected. A fair sky, gentle breezes and sparkling wavelets invite; and within a few brief hours the sky frowns, breezes become storms and the placid sea is transformed into a seething, swirling cauldron of angry waters, and the toll of the sea is paid in the lives of men and the agonies of their wives and children, 'For men must work, and women must weep.'"[1]

By August 29 an agitated *Daily News* was warning its readers that the toll of the storm was beginning to assume the nature of "a national disaster." "Feared Loss of Life May Exceed Half Hundred." The number of fatalities already exceeded thirty-five, twenty-three lost in Placentia Bay alone. Many other boats had not turned up, and the number of people injured and the loss of property "will make this storm unfortunately historic." News of the storm in Newfoundland was also filtering through to Nova Scotia (interestingly, even before the loss of their own schooners off Sable Island was being reported). A report out of St. John's in the *Halifax Herald* declared that between thirty and forty lives had been lost. The storm had swooped down over the coast without the usual warnings and caught the fishermen off guard. "It struck out of a cloudless sky and for hours a ninety mile gale blew while the sun shone brilliantly overhead." The localized nature of the storm tends to back up other reports that the weather had been fine that day on the Grand Banks. The *Bridgewater Bulletin* correctly observed that the whole story would not be told for some days, "so completely did the tempest demoralize the facilities for communication between the isolated ports of the Island."

Survivors, as usual, told harrowing tales. The schooner *Madonna Hayden* arrived in Harbour Buffett on August 30, having battled successfully with the storm. The vessel ran for Red Island in Placentia Bay and all was going well, "though seas were running mountains high," till what the captain, Thomas Hayden, termed "a foul bowler" broke over the schooner, smashing the main boom though it had been safely secured. One crew member was badly injured. The *J. M. Hayden*, James Hayden master, reached Spencer's Cove, also in Placentia Bay, in a disabled condition. The wind had changed direction suddenly, causing a tremendous sea to run, throwing the schooner on her beam ends and taking dories, trawls, and everything movable on deck into the sea. Two men were washed overboard. One of them grasped a rope and was quickly secured, according to the *Telegram* of August 30, while the other "performed the unique feat of grasping and clinging to the mast-head till the schooner righted," arriving safely back on deck though suffering some injuries.

Not all the men and vessels at sea that day were so lucky. The *John C. Loughlin*, a western boat of 21 tonnes owned by John Henry Loughlin, had only recently been built, at Creston near Marystown. The schooner and the seven men aboard were from Red Harbour, a small settlement on the eastern shore of the Burin Peninsula, and from nearby Flat Island. Carrie Brennan related how her husband Edward had come upon the wreckage of a schooner at Ship Cove, near Argentia: "As the ship rose and fell with the rhythm of the waves, a man tied or strapped to the rigging, seemed to be waving or signaling to him....As the stricken schooner listed in the wind, the man, hanging from the crosstrees by his right arm, would dip below the waves. The fishermen soon realized that he was dead... the schooner was the JOHN LOUGHLIN and the body was that of her captain."

On September 5 Ed Brennan wrote to the Loughlins, saying that he had been the first man to board the stricken vessel. He explained that the rescue crew had brought the captain's body to land, "clothed and coffined him...We thought there might be a man lashed to the wheel but it was only a suit of oil skins twisted around the wheel. It was impossible to do much as the craft was so much under water.... We took up...a work coat and a clothes bag marked J. Barrett Woody Island, but not a sign of a body....It is too bad all the brave men lost their lives in this gale. We know the crafts and so on can be replaced, but the poor men after a summer's toil it is sad indeed. May God comfort their friends and help them to bear the blow."

Winnie, the sister of the three Loughlin men who were lost, remembered years later that it had been "a beautiful dawning, no sea and a splendid day on the water...the gale flew at them seemingly from nowhere...so intense no one had time to get the drying fish off the rapidly disintegrating flakes. Fish stores and sheds collapsed; supplies and drying fish tumbled before the gale like autumn leaves along the ground." Captain Albert Loughlin's was the only body recovered from the wreck. The wives of both Albert and Charlie

Loughlin were expecting a child; their brother Fred had only recently been married. The *John C. Loughlin* was salvaged, renamed *Velma*, and fished again, but out of respect for the several widows it never again put in to Red Harbour.²

Like the *John C. Loughlin*, the four-dory schooner *Hilda Gertrude* was "running in the bay," bound for the fishing grounds off Cape St. Mary's, when the storm struck. The one body that was recovered was at first thought to be that of Captain Danny Cheeseman of Rushoon, Placentia Bay. However, it was soon discovered to be the body of crew member Oliver Dicks. On August 30 Malcolm Andrews of Baine Harbour wrote to the editor of the *St. John's Daily News* on behalf of Dicks's relatives, saying that the man's burial had been "accomplished decently…The brothers of the L[oyal].O[range]. Association met the corpse on the wharf and paraded, first to the home of the deceased, then back to the Church of England where the burial service was read by the Lay Reader, and then the remains were laid to rest, in God's Acre, beside those of his father, to await the Resurrection Morning." Oliver Dicks, only twenty-two, left behind a wife and two children, a widowed mother, and two sisters. The schooner *M and J Hayden*, which had been fishing nearby, discovered the *Hilda Gertrude* on her beam ends, with men clinging to her side, and attempted a rescue. One of the crew, George Norman, according to Robert Parsons in *Toll of the Sea*, saw his twin brother Michael on the *Hilda Gertrude* and leaned over the railing to shout something to him, but his last words to his doomed brother were lost in the gale.

On the other side of Placentia Bay, people were mourning the loss of the *Annie Healy*, a vessel that had been built at Fox Harbour near Argentia in 1900, named for the builder's daughter, and after a quarter century at sea had undergone a major refit during the previous winter. She was commanded by John Mullins. According to the *Telegram* of August 30, the *Annie Healy* could be seen from Argentia as she foundered. Seventy-three years later, in August 2000, the *Placentia Charter* reported that the residents of Fox Harbour and descendants of the men lost had unveiled a monument and a storyboard depicting the tragic event. The seven men on board "Big Annie," as she was known, were all married except the captain's seventeen-year-old son Michael, and they left behind six widows and thirty-four children. Descendants recalled that there was no such thing as welfare back then; people had to depend on themselves, their friends, and neighbours. After a parade around the harbour, a play by Darrell Duke called *Thursday's Storm: The Annie Healy Story* was performed, mainly by the residents of Fox Harbour.

Farther west along the south coast the *Effie May*, Captain Arthur Durnford, went down with six men. Two of them were brothers of the captain, and two others were his nephews. All of the men came from Rencontre West, an isolated outport about halfway between Harbour Breton and Burgeo. The schooner was on its way home, having had some repair work done at St. Pierre, and was never seen again. And far to the west, at Fox Roost near Isle aux Morts, as the *Daily News* reported on August 27, the *Vienna* with six men mainly from Burnt Islands was lost, as was the *Annie Jane* of Isle aux Morts, with four men, apparently on the way to St. Pierre.

The small schooner *Valena R.*, William G. Skinner master, belonging to Richard's Harbour and cleared through Pushthrough, a community near the mouth of Bay d'Espoir west of Hermitage, was discovered ashore at Morgan's Island near Lamaline with a lamp burning in the forecastle and nobody aboard, rather like the *Mary Celeste*. According to Robert Parsons in *Wind and Water*, two anchors were in place on the bow, a dory on the deck, and ten quintals of fish in the hold. Items on board suggested the vessel had recently been at St. Pierre and perhaps had been set adrift from the harbour there by the storm. Some of the men in Lamaline and nearby Allan's Island repaired the schooner and in the spring generously restored it to a surprised owner free of charge.

NEWFOUNDLAND VESSELS LOST ON AUGUST 25, 1927

ANNIE HEALY:
Captain John Mullins
Michael Mullins, 17, son of the captain
Patrick Bruce
Jack Foley, 57
John Kelly
James King
Charles Sampson

All the men were from Fox Harbour, Placentia Bay

EFFIE MAY:
Captain Arthur Durnford, 34
George Willie Durnford, brother of the captain
John Thomas Durnford, brother of the captain
Frank Durnford, 16, son of John Durnford
Garfield Durnford, 12, son of John Durnford
Benjamin Herritt

All the men were from Rencontre West

HILDA GERTRUDE:
Captain Danny Cheeseman, 30, Rushoon
Michael Hann, 22, Rushoon
Michael Norman, 22, Rushoon
Oliver Louis Dicks, 22, Baine Harbour
Patrick "Paddy" Gaulton, 22, St. Joseph's
Thomas Hawco, about 50, St. Joseph's
Thomas Keating, St. Joseph's
John Murphy, Parker's Cove

JOHN C. LOUGHLIN:
Captain William Albert Loughlin, 38, Red Harbour
Charles J. Loughlin, 22, Red Harbour
Frederick E. Loughlin, 29, Red Harbour
Herman Peach, 25, Red Harbour
Gordon Frampton, 24, Flat Island
Josiah Stacey 18, Woody Island
Josiah Barrett, 29, Woody Island[3]

VIENNA:
Captain John Chaulk
Robert Herritt
Freeman Organ
John P. Keeping
Thomas Keeping
George Strickland

All from Burnt Islands except for George Strickland, who came from nearby La Poile.

ANNIE JANE:
Captain Wilson Green, Isle aux Morts, apparently on the way to St. Pierre; four men lost

LORETTA:
Lost near Point Verde, just south of Placentia, with seven men, only one body recovered.

The *Daily News* of August 30 provided a list of the dead known so far: Fox Harbour, seven; Flat Island, seven; Rencontre, six; Burnt Islands, five; Rushoon, three; Patrick's Cove, two; Clattice Harbour, two; and one each from Baine Harbour, Red Harbour, St. Joseph's, La Poile, Humbermouth, and Belleoram, for a total of thirty-eight. The man from Humbermouth, near Corner Brook, was George Hayden, forty-nine, who at the height of the storm was blown off a wood boom and drowned at the Newfoundland Power and Paper Company. Like the list above, there are no doubt errors and omissions in the *Daily News*'s tally. Sources conflict, but it's clear, in any case, that more than forty fishermen died off the coast of Newfoundland in the August gale of 1927.

Cod is really the life of Newfoundland. You don't need a chart or a compass to find the Province; all you need is a good nose. The manners, customs, lore, tradishuns, laws, language, of Newfoundland all revolve about cod—leastways so us New Englanders think. But the funny part of it is, the people themselves seldom speak the word "cod." They allus call cod "fish." In an outport boarding-house I once heard an old Newfoundland sea-dog say, "If ye ain't got no fish, missus, then gi' us halibut!"

—Chelsea Fraser, *Heroes of the Sea* (1924)

Newfoundland. The big island. Ten thousand or so kilometres of coastline. Its history forever intertwined with the sea. As long as people have lived on the island, fish have been caught in the waters off its shores. Harassed by newcomers from Europe and Mi'kmaq from the mainland, vulnerable to European diseases and firearms, the Beothuk eventually disappeared: Shanawdithit, the last of her people, died at St. John's in 1829. About the time of John Cabot's transatlantic voyage in 1497, having discovered that the Grand Banks teemed with codfish, French and Portuguese fishermen began to exploit the Newfoundland fishing grounds. They were followed by the Spanish, with largely Basque crews, and in due course by the English. For centuries the Europeans sailed to the banks, fished for a few months, and returned home with their catch, the island being not much more than a "wharf," a convenient place to get fresh water and repair their vessels.

The English were at something of a disadvantage in this migratory fishery for they had no local source of the salt needed to preserve the fish until their ships reached their home ports. Access to the land thus became important, for wood was needed on shore to dry the fish in the sun. Their lightly salted, well-dried fish was preferred by many consumers. According to historian David Alexander, English West Country fishing boats made an "undignified annual rush" to claim seasonal property rights over the best fishing "rooms" (a "room" being the portion of shore on which the fisherman cured his catch and erected the necessary huts, stages, and flakes). In his history of the Atlantic provinces W. S. MacNutt described Newfoundland as "a great common for the seasonal use of English visitors." The island itself was not much valued. As an old West Country fishermen's rhyme put it: "If it were not for wood, water and fish, Newfoundland were not worth the rush."

The first fishing masters to arrive in each harbour in the spring, known as "fishing admirals," had first choice of the rooms, and were authorized to settle disputes among crews that arrived later. Thus effective jurisdiction was given to the "rough, customary justice of the fishing smack and the rum-keg court."[4] In time the fishing admirals were replaced by senior naval officers. Indeed, the fishery was considered the "nursery of seamen," essential for the maintenance of British sea power, for it trained thousands of men for the skills and toils of the sea, preparing them for the Royal Navy.

Given the nature of the migratory fishery, settlement was inevitably slow; permanent residents meant competition and interference with the fishermen's seasonal claims to land ownership. Regulations in the seventeenth century forbade English ships to carry settlers to Newfoundland or to leave behind any of their crews, but historian Keith Matthews and others have maintained that the laws prohibiting settlement were in fact short-lived and not successfully enforced. Planters and their servants from the southwest of England, from Devonshire, Dorset, Somerset, and Hampshire, began to settle permanently in the early 1600s. They built houses in the coves and inlets or on offshore islands, raised families, grew vegetables, and fished for their own use, but they also looked after the fishermen's rooms in winter and cut wood for use by the migratory fishermen in the summer.

There were attempts as well to promote settlement and encourage investment, sometimes with exaggerated accounts of the island's glories. In 1620 John Mason, in his *Briefe Discourse of the New-found-land*, wrote of the island's many resources, including fish: "But of all, the most admirable is the Sea, so diversified with severall sorts of Fishes abounding therein, the consideration whereof is readie to swallow up and drowne my senses not being able to comprehend or express the riches thereof....Cods [are] so thicke by the shoare that we heardlie have been able to row a Boate through them."[5]

Despite early efforts at settlement, notably John Guy's colony at Cuper's Cove (Cupids) and others at Ferryland and St. John's, the population in the seventeenth century remained sparse and scattered, and of little importance to the authorities compared to the fishery. According to Leslie Harris, one-time president of Memorial University, what emerged over the years was "an unplanned, unwanted colony, unrecognized as such until the 1820s"; the people who comprised the "hundreds of isolate little anarchies, clung like proverbial limpets to the rocks they had chosen as home."[6]

Meanwhile, the "merchant princes" in England raked in the profits. A lavish Georgian mansion, Upton House near the harbour in Poole, was built from the proceeds of the Spurrier Company's saltfish operations in Burin. En route to the fishing grounds West Country vessels called into Irish ports such as Waterford and Cork each spring to take on provisions such as salt meat, cheese, clothing, and beer, and eventually they carried not only supplies but Irish workers. Quakers George and William Penrose made so much money supplying salt pork to the migratory fishery that they started a glass factory and grew even richer manufacturing Waterford crystal.

By the early eighteenth century the settlers had begun to outnumber the seasonal fishermen. MacNutt claimed that by 1712 there were nearly three thousand English settlers along the harbours of the Avalon Peninsula. S. J. R. Noel, in *Politics in Newfoundland*, described the island in the eighteenth century "as a curious hybrid of the British Empire, being something more than a fishing station, something less than a colony." Small boats were heading out to the nearby fishing grounds from Newfoundland's many outports. The English vessels, meanwhile, were increasingly turning their attention to the offshore banks, and the catch, instead of being cured on shore, was salted down aboard ship, not dried, and taken back to England as "green fish." By the end of the century the transatlantic migratory fishery had been replaced by a resident fishery; the British fishery had gradually become a Newfoundland fishery. The shore-based salt cod fishery would dominate Newfoundland's economic history in the nineteenth century and continue to be the single most important source of employment and income well into the twentieth.

It was clear that the sea, and the fish within it, would be forever at the heart of Newfoundland life. "The sea clothes the island as with a garment," wrote J. D. Rogers in 1911 in *A Historical Geography of Newfoundland*, "and that garment contains the vital principle and soul of the national life.... To the Newfoundlander the land is a forest or a 'barren'; the sea a mine or harvest field, and on the foreshore the yield of the sea is prepared for market." In *La Géographie de Terre-Neuve* (Paris 1913) Robert Perret expressed a similar view of the island: "In Newfoundland as nowhere else can one be made to feel the contrast between a land that is infinitely silent, motionless, poor in vegetation, above all poor in its variety of living creatures, and a sea which harbors every form of life."[7]

※ ※ ※

Over the nineteenth century the fishing industry in Newfoundland evolved into several distinct branches, mainly in pursuit of cod. In the long-established shore-based fishery the fishermen caught, salted, and dried cod for sale to export markets in southern Europe, Brazil, and the West Indies. Operating from small, often isolated outports, the shore fishery, in which all members of the family participated, was especially active along the northeast coast of the island during late spring and summer when the cod moved close to shore following the caplin. The lightly salted cod, dried in the air on wooden flakes near the fishing stages, was considered to be the best in Newfoundland, for the weather in the northeast, usually windier and sunnier there in summer, was ideal for curing fish.

Each year in late spring hundreds of fishermen and their families would migrate "down north" to participate in the Labrador fishery, hoping to augment their earnings from the shore fishery. The weather being more severe and the shores barren, the season much shorter, the fish were dried on rocky beaches or salted on board and brought to Newfoundland for drying. Three kinds of fishermen operated the Labrador fishery: the "floaters," who lived aboard their schooners and fished from them; the "stationers," who went to a particular cove or harbour where they lived in tilts, or huts, and fished from small boats; and, finally, the "livyers," who were permanent settlers. The salt cod from the Labrador fishery became known as the notorious "Labrador cure"; heavily salted and of poor quality, according to George A. Rose in *Cod*, it fetched a lower price than the salted cod produced in the shore fishery. The latter fishery was supplemented as well by the spring seal hunt—"swiling" as the old-timers called it—which developed because of the rising demand for oils, reaching its peak in the 1840s with about 700,000 seals being killed annually.

In the early days the Newfoundland shore fishery was run by the West Country merchants, who sold the supplies needed in the spring and in the fall set the price to be paid for the salt cod. As the fishery was gradually taken over by local firms, mainly in St. John's, this manner of control survived in the form of a credit relationship known as the "truck system." Fishermen were advanced supplies such as food and fishing gear on credit in the spring and in the fall they sold their catch to the suppliers. If they had a credit balance, which was far from certain, payment was often made in goods rather than cash, always scarce in the outports. Not surprisingly, the merchants tended to charge as much as they could for the supplies and paid as little as possible for the fish. The merchants kept the fishing families going, providing them with a market for their fish and some kind of protection against the years when the fishing was poor, but the fishermen were kept in constant debt, in poverty or very near to it—indeed, almost a form of bondage.

A decade or so after the middle of the nineteenth century, two events having to do with fish are worthy of note. In 1867 Newfoundland's Maritime neighbours, Nova Scotia and New Brunswick, had joined together with Canada East (Quebec) and Canada West (Ontario) to form the Dominion of Canada. An intense debate took place in Newfoundland as to whether or not the island should join Confederation. The merchants were opposed, as were many Irish Catholics. One prominent anti-Confederate orator, Charles Fox

Bennett, warned against sending "your sons to die on the desert sands of Canada." A well-known song from the time sums up the feelings of many Newfoundlanders:

> Men, hurrah for our own native Isle, Newfoundland,
> Not a stranger shall hold one inch of her strand;
> Her face turns to Britain, her back to the Gulf,
> Come near at your peril, Canadian Wolf!

In his biography of Sir John A. Macdonald, *Nation Maker*, Richard Gwyn observed that a series of disastrous fishery failures in the mid-nineteenth century had put a third of the able-bodied population on relief, this at a time when 89 per cent of the work force was engaged in the fishery, and fish and seal accounted for 95 per cent of exports. Had an election been held in 1868, Gwyn contended, the island likely would have joined the union. However, by the election in 1869 an exceptional harvest of seals and cod had reversed public opinion. It would be three-quarters of a century before Newfoundlanders decided, albeit somewhat reluctantly, that their future lay with Canada.

The other important event concerns the banks fishery. Though some English vessels had fished on the offshore banks in the eighteenth century, based either in Europe or on Newfoundland, the early fishery had largely disappeared, in part because it was unable to compete with the French and American fishermen who were heavily subsidized by their governments. However, perhaps partly inspired by the arrival of Nova Scotia fishermen on the banks, the Newfoundland government in 1876 sought to revive the banks fishery by introducing bounties. Closest to the rich fishing grounds and in possession of plenty of bait, it was thought that Newfoundland, by rights, should be a major player in the industry. The subsidies for shipbuilding as well as outfitting, and several moderately successful fishing voyages, helped re-establish the banks fishing fleet. By 1880 there were 32 banking schooners with 432 crew members. The Newfoundland banks fishery reached its peak in both landings and number of men employed in 1889 with 330 vessels and 4,401 crew members, accounting for 20 per cent of all exports of dried cod. Thus, by the late nineteenth century the banks fishery had become the third cod fishery operating out of Newfoundland ports.

Initially St. John's was the main port, with sixty banking schooners by 1889, but many firms found the vessels expensive to operate and profits too low, and prices for salt cod were falling along with increased competition from Iceland, Norway, and France. During the 1890s the industry became centred on the south coast, especially in Fortune Bay and Placentia Bay, on either side of the Burin Peninsula. There, after 1900, the fishery was dominated by the merchants and independent vessel owners in the towns of Burin and Grand Bank. In 1914, out of a total of 105 vessels in the fleet, 104 were from the south coast. By the 1920s the fishery had left the northeast coast altogether, remaining in the south until replaced by the deep-sea dragger fleet in the 1940s and 1950s.[8] The last saltfish banker would make its final journey from Newfoundland to the banks in 1955.

The town of Grand Bank, on Fortune Bay, the port most closely associated with the schooner fishery on the Grand Banks, became known as "the bank fishing capital of Newfoundland." The town's prosperity began in 1881 when Samuel Harris, a local vessel owner and captain (who went to sea at the age of ten and skippered a vessel at the age of twenty-two), took his schooner the *George C. Harris* to the banks. The success of this venture helped launch a new era in the fishing industry. At one time there were as many as six shipyards in the town, and schooners were also purchased from New England and from Lunenburg, Lockeport, and Shelburne in Nova Scotia. Samuel Harris established a general merchandise and fish-exporting business in 1895. By 1926 he had owned and operated more than sixty schooners, fourteen of them patriotically named after First World War military heroes. Other prominent Grand Bank fishing entrepreneurs were George Abraham Buffett and Simeon and Daniel Tibbo. The town's architecture, like parts of Lunenburg and Gloucester, came to reflect the prosperity of the deep-sea fishery, with fine examples of Queen Anne–style sea captain's houses, topped by a widow's walk. The town bustled with activity when the fleet was in, with hundreds of fishermen preparing schooners for their spring trip or dispatching their catch. By 1910 Grand Bank was being referred to as the "Gloucester of the North."

Lunenburg vessels drying sails at Burin, Newfoundland, about 1920.

Grand Bank historian Robert Parsons in *Lost at Sea* described the hustle and bustle in the port as late as the 1920s: "This was the time of the great fleets of banking schooners, when every able bodied man could sling his duffel bag over his shoulder and go to sea as a doryman or as a foreign going seaman on the coasters and terns. Harbours were seemingly full of sailing ships; a forest of masts and spars. In the nineteen twenties proud citizens could boast that, 'in the spring just before the fishing season began, sure you could walk across our harbour stepping from one schooner to another.'" "Yet within twenty years," Parsons lamented, "the era of the sailing vessel passed and the South Coast harbours lay empty."

Burin, on the Placentia Bay side of the peninsula, was equally important in the banks fishery. The French and Basques had fished in the area in the early days and by the middle of the eighteenth century English Planters and fishermen were settling on the mainland, having abandoned earlier settlements on nearby islands. Robert Parsons noted in *Toll of the Sea* that captains and merchants "built fish drying premises and sent schooners by the score to reap the riches of the bountiful banks near Burin's doorstep." As in Lunenburg, the offshore fishery provided employment for hundreds of people on the Burin Peninsula and elsewhere, from the makers of sails, to blacksmiths, to clerks in stores, and the builders of dories and vessels. At Burin, wrote Garfield Fizzard in *Unto the Sea*, large flakes were constructed for the curing of fish, and at Grand Bank the beach was expanded for the same purpose.

A fishing vessel of considerable importance on the south coast was the western boat, known locally as the jackboat ("jack" meaning smaller or shorter in Newfoundland in this context). Small, square-sterned, two- to four-dory schooners that fished on the banks closer to home, they were the boats that suffered so badly in the August gale of 1927 while larger vessels fishing on the Grand Banks were spared. They were known as western boats by fishermen elsewhere on the island because, some said, they were built, owned, and captained by men on the south coast west of Cape Race, or west of St. John's.

Rear-Admiral H. F. Pullen in *Atlantic Schooners* described the "jack" schooners as "transom-sterned vessels...generally between 45 and 50 feet long, fore-and-aft rigged with a long bowsprit. The sail plan was quite simple, consisting of two jibs, a foresail and a mainsail. Topmasts were not fitted. The hull was designed to produce a graceful, fast and seaworthy vessel, well suited to the rough and tempestuous conditions to be found in Newfoundland waters." Most had the rudder "out-of-doors" (hung on the counter).

In *Beautiful Ladies of the Atlantic* Otto Kelland explained that these sturdy boats were built by men who had felled the timber themselves in the deep woods during late fall and winter, and at one time they were constructed in nearly every harbour from Bay Bulls to Port aux Basques. He also claimed that the skippers were noted sail carriers, some even reckless but never foolhardy. While the larger schooners stayed away for days at a stretch, if not weeks, the smaller vessels were rarely away from port for more than a few days at a time. According to Raoul Andersen in *Voyage to the Grand Banks*, Captain Arch Thornhill, after fishing out on the banks in 1918, continued to

A Newfoundland "jack" schooner or western boat, drawing by L.B. Jenson.

fish around Cape St. Mary's till the end of September. There were fifty or sixty jackboats fishing off St. Mary's that year, he observed; he often saw them anchored, jigging for squid. Thornhill maintained that there were hundreds of these western boats around Placentia Bay, St. Mary's Bay, Hermitage Bay, and some in Fortune Bay.

They fished off St. Pierre, St. Mary's, and Labrador, but only occasionally on the Grand Banks when fish were scarce on their home grounds.

Otto Kelland, who was born in 1904 in Lamaline on the Burin Peninsula and died only one month short of his hundredth birthday, accomplished many things in his long life, but he is perhaps best known for his well-loved folk song:

Take me back to my Western boat,
Let me fish off Cape St. Mary's
Where the hagdowns sail and the foghorns wail...
Let me feel my dory lift
To the broad Atlantic combers
Where the tide rips swirl and the wild ducks whirl
Where Old Neptune calls the numbers...⁹

A hagdown is a bird, a shearwater. The *Dictionary of Newfoundland English* tells us that hagdown is also a nickname for a man from Placentia Bay.

The deep-sea fishing season off Newfoundland began in late February or March when the vessels headed for the nearer banks such as Burgeo and St. Pierre, and later went on to the Misaine and Quero Banks. The main bait used was herring that had been caught during the winter and frozen naturally. By June caplin was being used to fish on the Grand Banks. Once anchored, the dories set out from the schooners in different directions. One day when the fog cleared a fisherman said it was "almost as if we were in a field of woods"; there were vessels in every direction, from Newfoundland, Nova Scotia, New England, and three-masted square riggers from France. In September many of the schooners sailed down north to the coast of Labrador for a fall trip, using squid as bait. Squid was also used as manure for potato and cabbage gardens, and some was dried as food for dogs in winter. Because the harbours on the south coast were generally free of ice, some fishermen were able to supplement their income by fishing on the nearby banks in winter. Many schooners were also employed in the coasting trade during the winter, sailing to Nova Scotia and Prince Edward Island especially for products not readily available in a land where the soil was thin, the growing season short.

The availability of bait was essential for the prosecution of the banks fishery. The herring fishing season began about the last week in November or early December, with small boats catching the fish in the bays, and later in the winter and in spring the harbours would be filled with vessels looking for bait, mainly from Gloucester and Lunenburg. Many local fishermen joined these schooners when they came into port. The vessels also needed ice to preserve the bait; cut from ponds during the winter, it was stored in ice houses and covered with sawdust for insulation until taken aboard. Since neither bait nor ice lasted very long, both foreign and local schooners frequently headed into towns like Aquaforte, Ferryland, and Cape Broyle on the east coast of the Avalon Peninsula, known as the Southern Shore, for fresh supplies.

The Newfoundland banks fishery was similar to that of Lunenburg and Gloucester, but there were differences. There was, for instance, greater reliance on the count system, in which a fisherman's earnings depended on the number of fish his dory caught, each fish being counted as it was heaved aboard the schooner. As social anthropologist Raoul Andersen noted, some men went to their graves "cursing the count," for dories were overloaded, sometimes to the edges of the gunwales, or dories went astray trying to find even more fish. Sometimes large fish were thrown away to make more room for small ones, to make a larger count.

Unlike the Nova Scotian and New England vessels, the Newfoundland schooners carried no dressing crews—the throaters, headers, and salters who remained on the vessel while the dorymen fished. When the deck became overrun with fish, likely after the second trip, the captain would usually hold back one dory crew to dress and salt the fish. The kedgie (or catchee, or flunkey as we met him in Lunenburg), whose usual job was to catch the painters as dories came alongside, as well as helping the cook and cleaning the mess off the deck, would cut the throats of the fish. One of the dorymen would gut, head, and pull out the livers. The captain took over the job of splitter while the second doryman would salt down the fish in the hold. Otto Kelland, who described this typical procedure in *Dories and Dorymen*, pointed out that methods of fishing naturally differed, depending on the captain.

Schooners both large and small used the bultow, a long line from which hundreds of baited hooks were suspended, the bait having been kept fresh in ice pounds on deck. These long trawls, reaching a kilometre or more in length, were set from the schooner or from dories off the larger schooners. Garfield Fizzard noted that in the 1860s some fishermen opposed to the bultow petitioned the House of Assembly to outlaw it: unlike the handline's hook, which was kept off the bottom, they argued, the trawl rested on the bottom and would catch the mother fish full of spawn, and lead to the destruction of the fishery.

In the Newfoundland bank fishery the merchants owned both the schooners and the fish (in Lunenburg shares in the bankers were sold widely, usually to local people). The skipper and crew worked for a share of the catch. Accounts were usually settled up before Christmas in a private meeting between the fisherman and the merchant, who typically provided no details about how many fish were caught or how much the fisherman's bait or food had cost. One old Grand Bank doryman, quoted by Andersen, recalled a bit of doggerel he had heard as a boy in Little Bay, Fortune Bay:

Get up early in the morning,
Go and bait your trawl.

> You scarce had time to light your pipe,
> When over your dories go.
> And make three runs a day,
> No matter how hard she blows.
> And the devil is in the merchant, boys.
> He just keeps us alive.

The sea provided a living, but it also, inevitably, took many lives. As Robert Parsons observed in *Lost at Sea*, of the thousands of vessels that were shipwrecked around the Newfoundland coast, hundreds were from communities in the south: "All along the coast, people were caught up in the agony or intense suspense of death or rescue from the sea, recovering bodies from the boiling surf or gleefully salvaging goods from wrecked schooners near the shore."

The dangers inherent in seafaring can be seen in the experience of Foote Brothers, a Grand Bank shipping and fishing business. Three schooners bearing the family name were lost within the space of twelve months. The *George Foote*, Captain Samuel Patten, with seventeen men on board, went down on the Grand Banks in the gale of August 22, 1892; the vessel appears to have dragged anchor and collided with another vessel, the *Cashier* of Lunenburg County. In the same storm the *Maggie Foote*, Captain Morgan Riggs, a small freight and trading schooner, foundered off Cape Race on the way to St. John's with a crew of five. Two Foote brothers were lost: George, who had been married only five months, and Clarence, who was planning to go to college in New Brunswick. One reason for the trip was to witness the destruction of St. John's, the capital and commercial centre of the country, after the great fire of July 8, the worst disaster in the city's history. Less than a year later, apparently in an August gale in 1893, the *Clarence T. Foote*, with Captain Richard Durnford and a crew of five, disappeared without trace on a trip from Grand Bank to Sydney in Cape Breton. Samuel Patten's wife Jane, left to raise three small boys, dreaded the thoughts of them sharing the fate of their father and moved to faraway Ontario.

A fishing schooner from Flat Island, Placentia Bay, the *Reason*, Captain Isaac Crann, also foundered in the August gale of 1892. Ninety-six-year-old Jessie Hale, in 1978, recalled in Rainer K. Baehre's *Outrageous Seas* how her brother Albert Joyce, nineteen, and others in the community were lost. Her stepfather had two brothers, three nephews, and a stepson on board. She called the men the "singing crew," for they always sang songs like "The Sailor's Farewell" with rich and vibrant voices as they weighed anchor. She said that she would never forget that day, for she had been wakened by the house shaking in the terrible storm. "I felt it was all wrong. When I was thoroughly awake, I realized why. This was the day my brother was expected home." Out of her window she saw that the roofs of several buildings had blown off, fences were down, and several boats in the harbour had drifted from their moorings. "Out to sea everything was white as far as the eye could reach.... The ocean was mad. It surged, boiled, foamed." Looking out to sea, a captain she knew pronounced: "No boat could bide afloat today." When night fell, the family came together and prayed. Ten men from Flat Island died that day, and the small community never completely recovered. For years afterwards, commenting on the fierceness of the weather, old fishermen would say: "Sure, it's blowing harder today than when the Cranns went down."

In the waning years of the nineteenth century the Newfoundland fishery was in decline. Some banks and merchant firms failed; public debt was growing, bankruptcy looming. The government attempted to diversify the economy by completing the narrow-gauge railway through the interior in 1898 and hoping to open up the country to agriculture and industrial activities, mining, and lumbering. But the hard times continued into the new century. Six out of ten people at work in the early years of the twentieth century were fishermen, according to historian Sean T. Cadigan, and many others prepared and handled fish products. Some merchant firms tried new products, such as Scotch-cured herring and a type of canned cod rather unattractively labelled "fish cheese." Innovations such as steam trawlers and cold storage plants were well known because of contact with the foreign fleets, but, unlike New England, Newfoundland firms could not count on a large domestic market for fresh fish. Captain John Lewis of Harbour Breton had installed a motor in his vessel, the *Metamora*, at the beginning of 1914, and demonstrated the value of motor-propelled vessels for the bank fishery, but most of his contemporaries regarded engines as an unnecessary luxury.[10] With prices for fish low and markets glutted, it was not the best of times for Newfoundland fishermen.

The island was embroiled in an August storm of a different kind in 1914, when Britain and her Empire declared war on Germany on August 4. As it did all over the world, the First World War brought hardship, changes, and sometimes good fortune. In Newfoundland's case the war led to unprecedented, if short-lived, prosperity in the fishing industry, for European competition was all but eliminated and the average price for fish between 1916 and 1919 was 65 per cent above the 1910–1914 average. Merchants and fishermen did well. There was no conscription in Newfoundland and temporarily well-off fishermen were reluctant to join up, but Newfoundland's death toll during the war was huge, per capita among the highest of the Allied countries. The dead came from all walks of life, but most, said George Rose in *Cod*, whether merchants' sons or men from the outports, had some connection to the fisheries.

Like the schooners of Nova Scotia and New England, Newfoundland vessels were the targets of German submarines. Robert Parsons maintains in *Survivors and Lost Heroes* that at least twenty Newfoundland schooners, either laden with fish bound for Europe or heading home with salt, were destroyed by enemy

U-boats. Usually they were boarded, valuables and food taken, the crew allowed to escape in dories, then bombs were placed aboard and set to explode; the small wooden vessels were not worth a torpedo. Abraham Thomas Cluett, captain of the schooner *Gladys M. Hollett*, with a cargo of herring out of Twillingate, was overtaken in August 1918 by U-156 not far off the Lunenburg County coast on the way to New York. The captain lost his clothes, his watch, and his nautical equipment, but especially upsetting was the fact that "the Hun commander" of the submarine claimed once to have commanded a Gloucester fishing vessel and that he had complete information on the Lunenburg fleet, a record of all the vessels, names of the skippers, numbers of men in the crew, tonnage, destinations, and so on. He said he had orders to sink the whole Lunenburg fleet, but had "no intention or desire of hurting or molesting the sailors." The men, with only the clothes they stood in, according to the *Lunenburg Progress-Enterprise* of September 11, 1918, took to their boats and landed on West Ironbound Island without mishap.

In the 1920s, the war finally over, the prosperous times came to an end. Prices for fish fell, and markets in Europe dried up as the fisheries there recovered. Norway, Iceland, and France reinvested in their fisheries, as did Britain; the Portuguese expanded on the Grand Banks, and France came back in force with steam trawlers. There was an abundance of salt fish for sale. More fish of the poorer quality was sold in the Caribbean and Brazil, but there was much poverty and destitution in the outports. The *Newfoundland Trade Review* declared: "Never, perhaps, in the whole history of the country have we faced more trying conditions in the great staple industry than during 1926." The south coast fishermen seem to have fared somewhat better, the *Canadian Fisherman* reporting in October 1926 that "around Fortune Bay the total catch was almost a record." In 1925 Harvey and Company purchased a steam trawler of 299 tonnes to add to its fleet operating out of Belleoram, but large-scale trawling was still some years away.

Times were clearly bad, and they would worsen over the next decade. At least the fishermen were now better organized and had a political voice through the Fishermen's Protective Union and its newspaper, the *Fishermen's Advocate*. Founded at Herring Neck in Notre Dame Bay in 1908 by William Coaker, the union established branches along the northeast coast, formed companies to provide goods and supplies to its members, and purchased fish for sale in international markets, challenging the oppressive credit system. The FPU was not much help to the banks fishermen, however. They were isolated on the south coast and their fish landings were only a small percentage of the total cod catch. According to Fred Winsor in "The Newfoundland Bank Fishery," they tended to support and seek fellowship and mutual assistance from denominational and fraternal organizations such as the Roman Catholic Star of the Sea Association, the Society of United Fishermen organized by the Church of England, the Masons, and the Loyal Orange Association.

Tough times in the fisheries sent many men to work in other industries. The *Canadian Fisherman* in May 1929 noted that fishermen were working in the iron ore mines at Bell Island, at pulp and paper mills, and in the forests near Grand Falls and Corner Brook. Many more men, wrote George Withers, were finding work in the fisheries of New England and Nova Scotia, for conditions appeared to be better on the foreign schooners: shorter working days, higher earnings, better provisions, separate crew to handle the fish once aboard, and no count system. The *Lunenburg Argus* of March 12, 1925, commented that "the old town by the sea is a hive of activity" for about two hundred Newfoundland fishermen had arrived to join the fishing fleet. A few years ago the county owned a fleet of 120 vessels, all manned by Lunenburg men; now the fleet was about half the size and many of the fishermen were Newfoundlanders. As the *Bridgewater Bulletin* remarked on January 11, 1927, some were saying that with fewer Lunenburg men to man the vessels it was necessary to find hands elsewhere. Very good substitutes could be found in

Newfoundland, for it had "many of the hardiest and best seamen in the world." In 1929 two Lunenburg schooners laid up in Burin over the winter for the convenience of the crew, all of whom lived nearby.

※ ※ ※

Almost two weeks after the Lunenburg schooners foundered off Sable Island on August 24, 1927, news of their loss began to filter through to Newfoundland. The first hint that the people of Newfoundland might be affected appeared in an alarming headline in the *St. John's Evening Telegram* of September 5: "Forty Lives, Chiefly Newfoundlanders, Feared Lost on Grand Banks."

Indeed, the scale of the disaster was just emerging in Nova Scotia. "Feared Schr. Joyce M. Smith Lost at Sea," read the banner headline across the front page of the *Lunenburg Progress-Enterprise* on September 7, 1927. "Sea Chest Seen As Evidence of Marine Tragedy." The item noted that sea-chests, in which fishermen packed their personal belongings, were not kept on deck but stowed below in the cabin or forecastle; to find a sea-chest drifting on the water implied that something serious had happened to the vessel. The *Joyce M. Smith* had left Lunenburg on June 8 on her summer trawling trip and had last been reported as sailing from Queensport, Guysborough County, on August 20. Now the steamer *Albertolite* had spotted detritus twenty-nine kilometres off Cranberry Light, Canso: "Passed a broken dory partly submerged, a sea chest floating and other wreckage." The gale had come on so suddenly, the report said, that "it caught many a seasoned mariner napping."

The first Newfoundlanders to die in the fearful August gale of 1927, then, were the twenty men

Joyce M. Smith, taken from a passing schooner just before the storm struck.

Captain Edward Maxner, Lunenburg.

Samuel Warren Jr., Salt Pond, Newfoundland.

aboard the Lunenburg schooner *Joyce M. Smith* when she went down off Sable Island on August 24. Apart from their three Lunenburg shipmates—Captain Edward Maxner, one of the veteran skippers of the Lunenburg fleet, his twenty-four-year-old son William, and John Peterson, the cook—the entire crew was from Newfoundland, mostly from the region around Burin and Marystown on the Burin Peninsula.

Small communities such as Salt Pond, with only sixteen families, were facing hard times. Money was scarce. Each family cultivated a small plot to grow vegetables, had a cow or two, a few sheep and hens, to supplement their meagre earnings. Consumption (tuberculosis) was rampant in the community, with some families being almost wiped out; Samuel Warren Sr., lost on the *Smith*, and his wife Rose Keating, had ten or eleven children, all but two of whom died of tuberculosis. And now the fury of the sea was taking its toll, for the men depended on the fishery for a living.

News of the disaster did not reach the Burin Peninsula for several weeks. Samuel Warren Jr., eighteen years old, left behind his parents and seven younger siblings; two other sisters had already died. When the schooner was in Burin for bait in June, Samuel gave to his family a formal photograph he'd had taken that spring in Lunenburg. Early in September Samuel's mother Catherine and his brother Norman walked from Salt Pond to Burin to barter butter and vegetables for other supplies. While in Holletts & Sons a customer asked if anyone had heard about the loss of the *Joyce M. Smith* with all hands. Some

George Burbridge, Epworth, Newfoundland.

of the people knew that Catherine Warren had a son, a brother-in-law, and two nephews on board. It must be another *Smith*, they said. But on passing through the village of Salmonier on the way home, a local resident asked if they'd heard about the *Smith*. Catherine became hysterical and had to be helped home. Several days later the nuns from the convent at Burin confirmed the news of the tragedy. Catherine gave birth to her last child on March 24, 1929, and named him Samuel in memory of his brother and uncle; the boy died of complications due to influenza before his second birthday.

A little is known about other members of the hapless crew. Benjamin Hamen and Thomas Farewell were said to be dorymates.

Farewell, from Creston South near Marystown, left behind a widow and five children. His nephew, Earl Crocker, also from Creston South, was survived by his parents and eleven siblings. James and Thomas Hodder, from the same small community, were cousins. John Whalen, from Fox Cove, who joined the vessel when she was in Burin taking on bait, left behind his wife and six children.

George Burbridge and his brother Charles were from Epworth, across the arm from Burin. Both men were engaged to be married at the end of the trip, Charles to the stepdaughter of shipmate John Pike. A daughter was born to George after his death; she was baptized Joyce in honour of her father and the schooner on which he lost his life. George Burbridge Sr., a widower, learned of his sons' deaths from a passerby while out for a walk in early September; he died shortly thereafter and it was widely believed that he died from grief.

Robert Cheeseman, in his late fifties, and his eighteen-year-old stepson Philip, who went as header on this trip, were from Lewin's Cove on the Burin Bay Arm. Philip's mother had been reluctant to let him go and tried to encourage him to leave the vessel when she put in to Burin for bait. After the loss Mrs. Cheeseman received medical treatment for depression, or melancholy as it was then called. She always regretted giving in, saying "I could have at least saved one of them."

John Pike of Burin Bay Arm, father of seven children and a stepdaughter, had joined the *Joyce M. Smith* at Burin in June. Once again it was several weeks before his death was known. One day in September the family was in the meadow spreading hay and a bird flew overhead. "Look, a crow!" said Mrs. Pike. "We're sure to hear bad news before the day is done." Shortly thereafter the clergyman walked towards them. Mrs. Pike's son Douglas was crewman on another vessel out of Lunenburg. Eight months pregnant, she gathered her children around her, looked towards the minister, and said: "I know he's the bearer of bad news. If it is your brother

Douglas, I'll die too, but if it is your father I suppose I will have to carry on somehow to raise the rest of you."

Archibald Keating, age twenty-one, a cousin of Samuel Warren Jr., was from Salt Pond. His brother Michael Keating had been lost at sea in 1925. Another brother, Charles, was lost in 1931, when the rum-runner *Wilson T.* went down in a storm near St. Pierre.

LOST ON THE *JOYCE M. SMITH*, AUGUST 24, 1927[11]

FROM LUNENBURG:
Captain Edward Henry Maxner, 55
William Maxner, 24, son of the captain
John Peterson, 68

FROM NEWFOUNDLAND:
Andrew Barnes, Harbour Mille, Fortune Bay
Fred Barnes, Harbour Mille, brother of Andrew
Philip Barnes, Fortune Bay
George "Dodd" Burbridge, 24, Epworth, Burin
Charles Burbridge, 31, Epworth, brother of George
Robert Cheeseman, late 50s, Lewin's Cove, Burin
Philip Hodder, 18, Burin Bay Arm, stepson of Robert Cheeseman
Samuel (Earl) Crocker, Creston South, nephew of Thomas Samuel Farewell
Arthur Dominick, Belleoram
Thomas Samuel Farewell, 35, Creston South
Benjamin Hamen, Creston South
Thomas R. Hodder, Rock Harbour, twelve children
James Joseph Hodder, 30, Creston South, cousin of Thomas Hodder
Archie Keating, 21, Salt Pond, Burin, related to the Warrens
John Samuel Pike, 42, Burin Bay Arm, seven children
Thomas Poole, Belleoram, five children
James Warren, Salt Pond, Burin
Samuel Warren Sr., Salt Pond, uncle of Samuel Warren Jr., James Warren, and Archie Keating
Samuel Warren Jr., 18, Salt Pond
John Whalen, 47, Fox Cove, six children

Harold Horwood, in his 1969 book *Newfoundland*, paid tribute to the men of the Burin Peninsula: "They and their fathers were always on intimate terms with death and the violence of nature, for these were the men who manned the banking schooners, sailing and rowing into winter gales…watching their dory-mates go under, clutching helplessly at trawl lines, or waiting in the eternal fog for dories that would never return. It was a hard, cruel, dangerous life, with limitless scope for resourcefulness and heroism, a life of small material rewards, filled with struggle and fierce pride and the bitter joy of hard accomplishment."

The forces of nature were not through with the people of the Burin Peninsula. On November 18, 1929, a tsunami, caused by a rare underwater earthquake on the southern edge of the Grand Banks, struck the southeastern coastline of the peninsula. Giant waves roared into the scattered coves and harbours at about forty kilometres an hour, lifting houses off their foundations, sweeping schooners and other vessels out to sea, destroying stages and flakes, damaging wharves and fish stores. The storm also tore up the seabed, likely contributing to poor catches in years to come. Property damage was estimated at about one million dollars. Worst of all, twenty-eight people were killed, and hundreds more were left homeless and destitute with winter approaching. Because the single telegraph line linking the peninsula with the rest of the island had been severed by a recent storm, it was three days before outsiders learned of the disaster and were able to send help.

Six years later disaster struck again. On August 25–26, 1935, a fierce storm struck suddenly out of the northeast, uprooting trees and destroying fishing stages, fences, fish stores, and telegraph poles. Vessels were driven ashore, dories and motorboats swept away, thus crippling the fishery. At least thirty-three men died at sea in the storm.

Perhaps the most moving story of the 1935 August gale concerns the Walsh family of Marystown. The wreck of the small fishing schooner *Annie Anita* washed ashore after the storm at Hazel Cove near St. Shotts on the southern end of the Avalon Peninsula. The vessel had been fishing out beyond Cape St. Mary's when the storm struck. The cabin was half filled with sand, which meant that the seas must have been breaking right from the bottom of the ocean. A small hand was found protruding up out of the sand. It belonged to Captain Patrick Walsh's twelve-year-old son Frankie, who, always subject to seasickness, had not wanted to go on the voyage; it was meant to be his final trip before going back to school. The body of his brother Jerome, age fourteen, was never found. A second body turned up on the *Annie Anita*, and it was assumed to be that of Captain Paddy. When the bodies were returned to Marystown, Lillian Walsh insisted that the man was not her husband, but the funeral and burial went ahead anyway. The next morning she managed to convince the priest that she could confirm the identity of the man by the scars on his body, and so the corpse was exhumed. The body turned out to be that of the mate, Thomas Reid, a cousin of Patrick Walsh who looked very like him. The Walsh's eldest son, James, twenty-one, had assumed his very first command that day as skipper on another small vessel; the *Mary Bernice* was found a day after the *Annie Anita*, bottom-up near Long Island at the top of Placentia Bay. There were no survivors.[12]

One man who did survive the storm was Michael Bruce, who, now in his eighties, recounted his adventure in the *Downhomer* of August 2002. He had been aboard the *James and Martha*, Captain James Bruce, and from the deck had helplessly watched the *Mary Bernice* be taken by the sea. Bruce compared living through such a storm and then returning to the sea to fish as something like a woman having a baby: "It's hard, I suppose, while you're at it, but once it's over you go back and do it again. That's just the way it was."

the GLOUCESTER CONNECTION

Gloucester is old, very old, and smells of fish. It has a happy-go-lucky trolley line, which smells of fish too; and a confectionery store where—it is a solemn fact—the ice cream tastes of fish.

—*New York Journal*, 1896

THE PORT OF GLOUCESTER, ON THE ATLANTIC COAST OF MASSACHUSETTS, was the deep-sea fishing capital of New England. Perched on Cape Ann, forty-five kilometres northeast of Boston, Gloucester sent men and vessels to the Grand Banks and beyond for generations. Its fine harbour was first visited by Europeans almost a quarter century after Sir Humphrey Gilbert arrived in St. John's, but because of its more southerly location and diverse resources, and because its milder climate allowed for year-round fishing, Gloucester developed more rapidly than its northern counterparts. Indeed, it became a model for them. Gloucester was the great innovator, in vessel design, the fishing trips to the banks, the use of dories and trawls, and in the building of large and impressive houses with wealth earned from the fishery. In the North Atlantic fisheries triangle, Lunenburg was sometimes referred to as "the Gloucester of Canada," and the town of Grand Bank on Newfoundland's Burin Peninsula as "the Gloucester of the North." These communities had long and intimate ties with New England, especially with the small city on Cape Ann, and by the 1920s a great many of the fishermen who sailed in Gloucester's vessels hailed from Nova Scotia and Newfoundland.

The deep-sea fishermen of Gloucester and the schooners they sailed in became the stuff of legend. George H. Procter, owner and editor of the *Cape Ann Weekly Advertiser*, who published *The Fishermen's Memorial and Record Book* in 1873, pointed out the dangers the fishermen faced. Theirs was "no holiday existence," he said, "but a continued grappling with the elements, a struggle for life, with storm and old ocean in its anger to meet," but "with pluck and daring" they wrung success "from the very verge of the grave." In 1924 in *Heroes of the Sea*, Chelsea Fraser, who wrote books for boys in the 1920s and 1930s, wrote of "the stalwart framed, stout-hearted Gloucester fishermen" who sailed to the banks: "None but the very toughest can stand such a strain"; these humble workers of "monumental modesty," who spent their lives "within reach of the sea's maddest tossings and within hearing of the sea's maddest roarings," were nothing less than "Supermen."

In *The Masts of Gloucester: Recollections of a Fisherman*, Raymond McFarland in 1937 claimed "the real glory of Gloucester is her masts and men. Far above the odors and wrack of the wharves rise the slender masts of ships, swaying gently under the impulse of throbbing tides.... These ships were alive; sails fluttered from their spars; pennants crowned their topmasts; they lived, and labored, and died in the service of mankind; they were traffickers of the deep, seasoned by storms and honorable by successful voyages to tempestuous banks and the ports of the world....Between the call of the sea and the silent lure of tall masts the imagination of youth runs high. These ships are the pride of America."

More recently, too, the Gloucester fisherman has been seen as living a special kind of life. "Among the endless legends of man's struggle with the sea none is more magnificently moving than that of the fishermen hailing from Gloucester, Massachusetts," wrote Sterling Hayden, who became an actor noted for his great height and blond beauty. "*Gloucestermen*," he maintained in 1983, referring to both men and vessels, was "a word long held in something close to sheer awe by generations of seafaring and literate folk the world over." The schooners that were sent to "the eternal sea" were "almost the mirror image of that great brawling young giant of a nation known as the *USA*. No chauvinism here. The ships speak for themselves."[1]

In *Gloucester on the Wind: America's Greatest Fishing Port in the Days of Sail* Joseph Garland, Gloucester's finest historian, referred in 1995 to the "romanticized 'glory days' of the windships," when, in the late nineteenth century, "upwards of four hundred of Gloucester's uniquely beautiful two-masted schooners sailed forth to the northwest Atlantic fishing grounds from Virginia to Greenland." "Nowhere in the Western Hemisphere," he asserted, "was that reckless romance with wind and wave pursued more fruitfully and fatefully."

Sebastian Junger, in 1997, in *The Perfect Storm* declared that "If the fishermen lived hard, it was no doubt because they died hard as well," noting that over the centuries some ten thousand Gloucestermen had been lost at sea, far more than had died in all the country's wars. "Sometimes," he wrote, "a storm would hit the Grand Banks and half a dozen ships would go down, a hundred men lost overnight.... On more than one occasion, Newfoundlanders woke up to find their beaches strewn with bodies." Mark Kurlansky, too, in *The Last Fish Tale*, in 2008 saw the Gloucester fishermen as a breed apart: "Gloucester was populated by simple men with modest incomes and without affectations, but who walked with an aristocrat's belief that they were a special breed—a breed that was never curbed by fear and knew things that other men missed."

It's not surprising that Americans would make much of the virtues and heroics of their deep-sea fishermen, but it was the Englishman Rudyard Kipling, in his 1897 novel *Captains Courageous*, who immortalized the breed. Taking a break from the glories of the British Empire and the "white man's burden," Kipling tells the story of Harvey Cheyne—a repulsive, spoiled brat of a rich kid who, having fallen overboard from an ocean liner, is miraculously saved by a doryman and trapped for several months aboard the schooner *We're Here*, where he learns to be not only a fisherman but, ultimately, a man, or at least "A Banker—full-blooded Banker." Kipling describes with admirable accuracy the routine of life and work on a fishing schooner, the hardship, the food, the back-breaking work, the camaraderie, and the telling of tales. His one brief trip aboard a real schooner, however, nowhere near the Grand Banks, didn't go at all well; seasick in his bunk, he told a crewman: "Kindly inform the captain I would prefer to lie and meditate."[2]

Kipling's novel was a great success, as was the 1937 film based on the book, but it was James Brendan Connolly, in his many books and articles in the popular magazines of the day, who, many believed, gave a more accurate, if romanticized, picture of the fishermen's lives. Frederick William Wallace, who thought Kipling made "many mistakes," in an article in the *Canadian Fisheries Manual* in 1931 called "The Banksmen" praised Connolly as "the only one who really knows his subject" because he had sailed to the fishing grounds with his heroes. In 1927 Connolly confessed in *The Book of the Gloucester Fishermen* that in his enthusiastic moments he rated "these bank fishermen of ours as the greatest sailormen who ever lived."

THE GLOUCESTER CONNECTION

To him, the fishermen of Gloucester were heroes, especially the master mariners: "Without them, the city, quite simply, would not exist." In their heyday, he avowed, "the superb schooners of the New England fleets were the wonders of the maritime world...the fastest and most weatherly fore-and-aft rigged vessels of all time." As for the men in the dories, their stories tell of "human fortitude and courage almost beyond shore-going belief." But they were not to be pitied. "Toil is theirs, and suffering and peril: but it is men's work—no boy's or woman's or half-made creature's,—it is men's work."

For Connolly the Gloucestermen were a gallant, superior breed of men. A review of *Out of Gloucester*, his first novel published in 1902, declared: "Here is romance—true romance, whose heroes are intensely alive, their antagonists the elemental forces, the scenes of their daring the open ocean or the harbors of the North." In a citation presented when Connolly was awarded an honorary degree from Fordham University in 1948, Ernest Cummings Marriner said: "He discovered the placid Gloucester fisherman, to whom adventure was commonplace, and with vivid, salty phrase made him into a legend, his ship now smothered in sea-spume, now racing home on its rail and the crew sitting out on the keel."

In 1983 in *Down to the Sea: The Fishing Schooners of Gloucester*, Joseph Garland wrote that Connolly's tales "exalted the stout hearts of the fishermen (with a happily Celtic bias, it must be admitted), personalized and glorified the vessels and captured eloquently the excitement of plowing along through the crested seas with every stitch of canvas aloft, every sheet straining, every pulse pounding and the decks awash to the wheelhouse." Many feature writers, he asserted, "discovered and rediscovered the glamour of tall ships a'sailing and the pungent romance of busy wharves and quaint streets, but it was Kipling and Connolly, the Anglophile and the Anglophobe, who really invented the rest of the world's image of Gloucester, an image that with time and the passing of sail ossified into stereotype."

The Gloucester fishermen, then, were, in the eyes of many, somewhat larger than life.

❦ ❦ ❦

In 1606 Samuel de Champlain, on an expedition to what was then known as Norumbega led by Jean de Biencourt, sieur de Poutrincourt, entered a handsome and excellent harbour which they named Beauport. Hundreds of Aboriginals gathered to greet the strangers, though they spat out the grape juice given to them by the French, thinking it might be poison. The next day the French were busy washing their dirty socks on the beach when they were approached by a group of people with weapons in hand. They began to dance, and Champlain asked them to dance some more. But then Poutrincourt marched towards them with a file of musketeers, and they scattered in all directions.[3] Thus began the history of Gloucester.

In 1623, just three years after the Pilgrims landed at Plymouth, fourteen Englishmen spent the winter on Cape Ann, setting up a fishing station in the harbour. By this time there was no reason to fear the Native people, for European diseases, from which they had no natural immunity, had wiped out the vast majority of them. J. B. Connolly, in *The Book of the Gloucester Fishermen*, related that "a divine" came among the early settlers and proclaimed: "Remember, brethren, that you journeyed here to save your souls." "And to ketch fish!" cried out one of the brethren, springing to his feet. According to Joseph Garland in *Down to the Sea* the original settlers came from Dorchester in Dorset, and then from Gloucester, some fifty kilometres up the River Severn from Bristol. By the 1630s West Country merchants had built stages, boatyards, and taverns, all essential for a permanent fishery. By the middle of the century the first boats were fishing on the banks off Nova Scotia and on to the Grand Banks. A trade developed with the southern colonies and the British West Indies, including the provision of the poorer grades of dried cod for

THE GLOUCESTER CONNECTION

African slaves on the sugar plantations. The vessels returned with molasses, tobacco, cotton, and salt. The best fish were sold in Spain in exchange for wine, fruit, iron, and coal. Fishing communities were also established at Marblehead, Salem, the Penobscot Bay area in Maine, and in other settlements along the coast.

By the middle of the eighteenth century Gloucester was home to a sizeable fleet. The first schooner had been built and launched from Eastern Point in 1713 by a man named Andrew Robinson. By 1740 about seventy offshore vessels were handlining in northern waters, most making two trips a year. Edmund Burke, in the British House of Commons in 1775, marvelled at the accomplishments of "this recent people": "Pray, sir, what in the world is equal to it? Pass by the other parts,

Gloucester harbour in the 1880s.

and look at the manner in which the people of New England have of late carried on their fisheries....No sea but what is vexed by their fisheries, no climate that is not witness to their toil." Mark Kurlansky, in *Cod*, noted that by the eighteenth century codfish had lifted New England from "a distant colony of starving settlers to an international commercial power," and Massachusetts had "elevated cod from commodity to fetish," with cod appearing on official crests, coins, and stamps.

The New Englanders did not fish for cod alone. From the early 1830s to the late 1860s, for instance, they maintained a large mackerel fishing fleet in the "The Bay," as they called the Gulf of St. Lawrence. Also in the 1830s, two centuries after the arrival of the first settlers, the Gloucester captains finally began to take their vessels for cod and halibut to Georges Bank, the closest major fishing grounds, about 305 kilometres east-southeast of Cape Ann, and according to Connolly in *The Port of Gloucester* "the most prolific breeding ground for fish in all the world, but also the most dangerous" because of the shoal waters and strong running tides. About 1850 the Gloucester fishermen also began fishing for herring, off Newfoundland, mainly to be used as bait.

For two centuries, Connolly maintained, Gloucestermen did their fishing from the deck of the vessel and by means of single lines. It wasn't until the third quarter of the nineteenth century that they began to take notice of the trawling methods of the French on the Grand Banks. George Procter claimed that trawl fishing was introduced about 1861, adding another peril to the fisheries, namely the many dangers associated with fishing from dories. About the same time, the mackerel fishery was moving from handline to seine netting.

As the nineteenth century progressed, the town increasingly prospered as a fishing port, according to historian John N. Morris in *Alone at Sea: Gloucester in the Age of the Dorymen*. Along with wharves, storehouses, marine railways, and new firms to outfit the vessels, stores, banks, and insurance companies came into being, as well as saloons. By the late 1860s Gloucester had reached its peak, and it would remain America's premier fishing port for many years to come. In 1873 the town became a city. Cod remained the basic staple in the Gloucester fishery, in 1910 still making up over half of the port's salted fish products. But it was in the 1890s that the salt cod fishery began to decline. The fleet of 339 vessels in 1888 had dropped to an average of 60 vessels yearly by 1901. (By contrast, the Lunenburg fleet in 1888 numbered 193 vessels, and in the early 1900s Lunenburg still outfitted 140 or so schooners.) In the early years of the twentieth century the catch continued to decline, as did the number of large outfitters, until in 1906 four of Gloucester's largest salt cod processing firms merged to form one big company, Gorton-Pew Fisheries. There was also increasingly less demand for salt cod as fresh and frozen fish became more popular, a trade that was largely in the hands of Boston dealers.

As in the Nova Scotia and Newfoundland fisheries, Gloucester suffered great losses over the years. An anonymous reporter wrote in 1876 that the "history of the Gloucester fisheries has been written in tears." In *The Last Fish Tale*, Mark Kurlansky told of an August gale in 1635 in which a small boat, the *Watch and Wait*, sailing from Ipswich around Cape Ann on the way to Marblehead, smashed onto the rocks taking twenty-three people to their deaths, including ten children. "The sea would take thousands upon thousands of Gloucestermen," he remarked, "and the resulting pain shaped the town of Gloucester as much as did the dense, gray-spotted granite or the fine, white-flaked cod."

Many Gloucester fishermen died in the Yankee Gale of October 3, 1851, named for the many American fishing schooners that were wrecked in the Gulf of St. Lawrence off Prince Edward Island. Under the Fisheries Convention of 1818, signed between Great Britain and the United States, the Americans were allowed to fish in the gulf but they were not supposed to be within a three-mile limit except in search of wood, water, and shelter. The fish, however, did not respect international boundaries—nor did their pursuers. When the storm came on, the vessels were not blown out to sea, and eventually they were trapped by the winds and many were driven onto the beaches and rocky headlands. Some capsized at sea or were run down by other vessels. According to historian

Edward MacDonald, Gloucester apparently had 140 fishing vessels and 790 men in the gulf at the time of the storm, and 9 vessels from that port were lost. "In memory," he wrote, "storms are not measured by the force of their winds or the amount of precipitation, but by their greed for life.... The carnage almost defied belief. Debris was strewn along the coast in grisly windrows: wrecks and parts of wrecks, masts, sails, rigging, ship's timbers, barrels of mackerel, sodden clothing, boots, boxes. And bodies.... Some of the victims were still lashed to parts of their vessels. Most had been stripped naked by the violence of the seas."

There was talk in New England of divine retribution for, contrary to the custom in the past, the fleet had been fishing on Sundays.

Stories from the Yankee Gale were not all doom and gloom. One schooner, the *Rival*, left Gloucester with "a musical crew": violinists, guitarists, pianists, "violincellists" were on board, even a "dancist" and a "singist," according to George H. Procter's *The Fishermen's Memorial and Record Book* (1873). The black cook "also manifested a musical taste," but a fellow seaman meanly greased his fiddle bow the first night out and he left the ship at the earliest opportunity. In the fearful gale of October 3, laden with mackerel and after a narrow escape from foundering, the vessel was driven ashore. A kind-hearted Island farmer took in the musical fishermen, and in return they provided an entertainment at his house. It went so well that the captain and one of the men, John Jay Watson, who later became a celebrated violinist, decided to go on tour to better their finances. First they played in Charlottetown to a crowded house, then on to Saint John and Portland. They gave a final triumphant concert in Gloucester after arriving home.

More Gloucester fishermen were lost in the great storm of August 1873, but 1879 proved to be the most disastrous year for the Gloucester fisheries, with 249 fishermen lost over the year, in twenty-nine vessels. Thirteen vessels and 143 men went to the bottom on Georges Bank in a single gale on the night of February 20–21.

There were other storms, other disasters, too numerous to mention. Some idea of the extent of the losses in the hard later years of the nineteenth century can be seen by the fact that 1900 was considered "a year of extreme leniency": only eight vessels and fifty-four men were lost. And in 1910 only one vessel went down, with twenty-five men, the smallest number with one exception for nearly half a century.[4] W. Jeffrey Bolster in *The Mortal Sea* contended that fishing made coal mining look safe. In the twenty-four years between 1866 and 1890, he noted, more than 380 schooners and 2,450 men from Gloucester were lost at sea—"a chilling average of more than 100 fatalities per year from that single occupation in a town of between 15,000 and 16,000 residents."

Like the women of Lunenburg watching for vessels passing the Battery Point lighthouse, families in Gloucester kept their eyes fixed on the harbour, straining to see if schooners rounding Eastern Point were returning with their flag at half-mast. In *Down to the Sea*, Joseph Garland wrote of all the various types of vessels that had disappeared over time, packets, clippers, merchantmen, whalers and coasters, salt carriers: "Of all this ghostly armada, none held on longer or paid off more handsomely at a more horrible cost in lives and living than the Gloucester fishing schooner. No class of vessels in history, possibly, served its home port in so paradoxical a role as both servant and master. The men and the schooners of Gloucester, in the days when she sent forth more of both to bring back more fish than any other port in the world, were in the eyes of the world as one. *Gloucestermen* they were called with wonder, schooners and men, so close was the symbiosis—some might say the deadly embrace—of town and sea."

For generations the fishermen of New England had been sailing north, their schooners a common sight in the harbours of the Maritimes and Newfoundland. The Gloucester mackerel vessels fishing in the Gulf of St. Lawrence in the nineteenth century, for instance, called in at Port Hood on the west coast of Cape Breton for bait, ice, barrels, fresh vegetables, and other supplies, and they also took on many local men as crew. American vessels frequented Newfoundland ports such as Cape Broyle on the Avalon Peninsula, Burin and Grand Bank on the south coast, and the Bay of Islands in the west. Canso, on the Nova Scotia mainland, located on the route to "The Bay," as well as being the closest Nova Scotia port to the banks fishing grounds, was also a favourite stopping place. And many Gloucester vessels dropped into the South Shore town of Shelburne for supplies, especially the popular Shelburne dories, their success based on the local invention of a metal clip for joining the floor futtocks to those of the sides, and the fact that at about twenty-five dollars apiece they were cheap to purchase. There were also shipwrights, sailmakers, and riggers in town, if needed.[5]

The New Englanders came into the northern ports for bait and supplies but also for repairs and sometimes for medical treatment. Since they often had to hang around for bait, there was time to meet the local people, to party, perhaps to have a drink or two, or three. The Americans were particularly welcome in Newfoundland because they brought cash into the local economy; in his thesis "Our Yankee Cousins" historian W. G. Reeves claimed that American gold became part of local folklore, for large amounts of it was supposedly hoarded by fishermen in improbable places. From the visitors in port, and from sailing on New England vessels, the Newfoundlanders learned much about fishing on the banks. Judge D. W. Prowse commented: "Our people are quick to copy anything. Fortune Bay has admirably copied the best American models in vessels, boats, and banking outfits."

The relationship was not always rosy, however, for the Americans sometimes fished too aggressively in northern waters. In January 1878 four of the twenty-two schooners from New England that had arrived in Long Harbour, Fortune Bay, began to seine for herring rather than buying from the locals. Fearing their livelihood threatened, and objecting to fishing on a Sunday, a crowd of about two hundred men went forth in boats, "making warlike demonstrations and using threatening language"; having seized the American seines, they released the fish and destroyed the nets. Captain Solomon Jacobs on the schooner *Moses Adams* had foreseen trouble and had armed his men with pistols, thus managing to hold the crowd at bay, eventually sailing away with a partial cargo and a reluctant fisheries officer who was dropped off at St. Pierre. The incident became known as "the Fortune Bay riot," though in Gloucester Captain Jacobs was known as the hero of "the Fortune Bay outrage." The Americans were eventually compensated for their losses.

From the early years of the nineteenth century more and more British North Americans had been taken on as crew on New England vessels, and many of them moved permanently to "the Boston states." John N. Morris noted in *Alone at Sea* that before the mid-1840s the schooners had been crewed by the men of Cape Ann, men of English roots who lived in the harbour area in communities like Annisquam, Lanes Cove, and Rockport. By mid-century fathers were tending to discourage their sons from fishing, urging them to take on less arduous and less dangerous ways of making a living, especially in winter. Old Gloucester names were disappearing from ships' papers, except as owners, making room for "brawny newcomers" from outside.

Dependence on foreign fishermen grew ever greater in the early 1860s as local fishermen took part in the Civil War. "[W]hen the town ran short of natives," Joseph Garland commented in *Down to the Sea*, "it filled their ranks with imports from seafaring nations on both sides of the Atlantic to a point where, in the 1890s, so they claimed, most of the skippers in the fleet were Bluenoses from the Canadian Maritimes, Newfies and Portagees from the Azores, with an Irishman and a Swede or two thrown in." "In fact, as time went on," wrote Morris about the first decade of the twentieth century, "it was only the seasonal migrants from Nova Scotia who kept the salt-bankers at sea."

The exodus of fishermen to New England left a shortage of the best in the countries they were leaving. As early as 1837 a complaint had been heard in the Nova Scotia legislative assembly: "The youth of the province are daily quitting the fishing stations, and seeking employment on board United States vessels, conducting them to the best fishing grounds, carrying on trade and traffic for their new employees with the inhabitants, and injuring their native country by defrauding its revenue, diminishing the operative class, and leaving the aged and infirm to burthen the community they have forsaken and deserted." The *Port Hood Eastern Beacon* wrote in 1879 of "the weekly leave-taking of so many Cape Bretoners for the United States by every Boston-bound steamer, as well as by Gloucester and Boston sailing vessels." Gloucester vessels often left home with a skeleton crew and picked up men in ports along the way.

In the Canadian House of Commons in 1888 a Conservative Member of Parliament from Shelburne County complained of the "very great exodus," saying that it was almost impossible to get the labourers he required. Many men went away in the spring and came back in the autumn with a good suit of clothes, but instead of bringing back money they lived on the old people during the winter. The *Shelburne Budget* of February 1888 was deeply offended by this "insolent remark...a more outrageous libel was never uttered." The men went south, the newspaper said, because they made more money there and were paid in cash as soon as their fish were weighed, unlike at home where the fishermen were generally paid only after the cured fish had been marketed, a practice that could take many months. A captain in nearby Lockeport in 1904, trying to find a crew for his fishing vessel, lamented that "[a]ll the men around here are talking Gloucester."[6] And in May 1927, in its Newfoundland news, the *Canadian Fisherman* reported that fishermen were leaving in droves; the Red Cross boats to New York, Boston, and Philadelphia on every outward trip that spring were taking hundreds of young fishermen south.

The Americans were not surprised that men from "the Provinces" would want to come to work on their vessels. George Procter in 1873 felt that "these men, as a class," were "naturally fitted for the business." Born and raised by the sea, mostly of poor parents, they had to earn their own living at an early age, and fishing was about the only occupation available to them. It was not strange that they would want to sail on better equipped vessels, with opportunities to get ahead, to become masters, even owners. In short, they "caught the inspiration of the Yankee fishermen." J. B. Connolly, much later, allowed as to how the "[p]rovincial fishermen had always been a skillful, hardy, courageous company. Young men went from the Maritime Provinces to fish out of Gloucester, and back home they would come with great tales of Gloucester vessels, of the big money to be made there, of the greater glory of the life altogether. The dream of many a young Provincial was to be some day fishing out of Gloucester, perhaps some day to command one of her great vessels. The dream came true for many of them. Some of the greatest skippers that Gloucester ever knew hailed originally from the Canadian provinces." And, as Joseph Garland observed, Gloucester, having run short on native talent, "depended to a very large extent for the manning and the mastering of its great fleet of schooners on 'whitewashed Yanks'—down-homers as we call 'em, Nova Scotians and Newfoundlanders, bluenoses and herring chokers, master mariners attracted to the 'Boston states' by a better deal fishing all around."

There was, said W. G. Reeves in "Our Yankee Cousins," sometimes an element of condescension towards the lesser breeds from the north. The *Gloucester News*, referring to the people of the Bay of Islands, declared that it was "a great day for the natives when the Gloucester men educated them to more modern times." Like Connolly, Raymond McFarland wrote positive portraits of the people of Newfoundland, but in one of his "true stories" he described the loutishness of a Newfoundlander with whom he had sailed, saying that such behaviour was understandable from a man "brought up on codheads and herring bones."

It's difficult, if not impossible, to be precise about the number of Maritimers and Newfoundlanders who moved to New England over the years. They were joining the multitudes pouring into the United States in the late nineteenth and early twentieth centuries, and many moved without much notice being taken of them. Others went south only temporarily, for work. Reeves claims that the longest running and largest migration of Newfoundlanders were the numerous fishermen who went to join the Gloucester fleet. Most who eventually went on to be captains tended to become naturalized, to marry, and to settle down in the various ports, but a great many of the crew chose to continue living in Newfoundland and to migrate annually.

Nova Scotians emigrated to New England from all over the province. Frederick William Wallace in "The Banksmen" noted that the men of Shelburne, Yarmouth, and Digby Counties were well known as offshore fishermen who gravitated between home in Nova Scotia and Boston and Gloucester, fishing in the fleets of those ports. He also maintained that the men of Lunenburg County usually sailed in their own vessels and "were never very much for fishing out of other ports." In *Markland or Nova Scotia* in 1903, Robert R. McLeod had made the same point, observing that Lunenburgers were "a 'home' people, and few, if any, of our fishermen go from the county to fish in foreign vessels." But some emigration from the county certainly did take place, especially in hard times, and notably during the period of stagnation in the decade before the First World War. On December 22, 1925, the *Bridgewater Bulletin* confirmed that the "tried and tested of the Lunenburg fishing crews" were "leaving for Gloucester in late years."

Statistics relating to disasters at sea give some idea of the "mongrel" nature of the Gloucester fleet as well as the extent of its dominance by British North Americans. In 1894 and the first month of 1895, for example, of the men who died in the Gloucester fishery 35 were from mainland Nova Scotia, 21 from Newfoundland, 15 from Sweden, 14 from Cape Breton, 9 from Portugal, 8 from Saint-Pierre and Miquelon, 6 from Iceland, 4 from Norway, 3 from the United States, 3 from Finland, 2 from Ireland, and 1 each from Germany and Italy.[7] A list of crew members in the Gloucester fleet in 1917, apparently made because local companies were complaining that it was cheaper to provision in Canada, is also instructive: 897 Canadians, 754 Americans, 237 Newfoundlanders, 137 Portuguese, 17 Scandinavians, 14 French, 3 Italians, 2 Irish, and 1 man each from Spain, Russia, and Germany.[8]

"Fishermen from less developed societies," Reeves commented, "were thus out front in facing the industry's ever present dangers." The *Canadian Fisherman* in March 1914 noted that twenty-six men were lost in the fleet in 1913, sixteen of them from Nova Scotia and four from Newfoundland; in the May 1916 issue the journal noted that in 1915 of the thirty-one men who "paid the toll to Old Ocean for the finny spoil they wrested from her waters," twenty were from Canada, four from Newfoundland, and seven from other countries. At the end of each fishing season the *North Sydney Herald* reprinted from the *Cape Ann Weekly Advertiser* of Gloucester a list recording the Cape Bretoners lost in the fleet that year, and the *Canadian Fisherman* reprinted the toll enumerated in the *Gloucester Times*. "When the lists of fishermen lost in the fishing fleets out of Gloucester are published annually, it is practically an obituary list of Canadians and Newfoundlanders," declared the *Canadian Fisherman* in 1916. "The mourning for the lost men is not done in American homes, but in the villages and hamlets of Canada and Newfoundland."

A measure of the impact Maritimers and Newfoundlanders had on the Gloucester fishing fleet can be seen by the number of them who became captains of some renown. At Yarmouth in the early twentieth century F. W. Wallace discovered that the majority of the masters of the American vessels in port were native-born Nova Scotians, and a considerable percentage of the crews were Bluenosers as well. "The international boundary between the New England States and the Maritime Provinces," he wrote in his autobiography *Roving Fisherman*, "appeared to be very thinly drawn insofar as the deep-sea fisheries was concerned."

Perhaps the most famous of the Gloucester schooner skippers was Captain Solomon Jacobs, "king of the mackerel seiners," born in Twillingate, Newfoundland, in 1847. Tall, strong, intensely competitive and innovative, according to Joseph Garland in *Gloucester on the Wind*, he rose rapidly to the top. He made and lost fortunes and sailed as far as Ireland in one direction and the north Pacific in the other. It was Jacobs who launched the first engine-powered schooner, the *Helen Miller Gould*, in 1900. In Newfoundland, said Reeves, he was a local hero, "Our Sol Jacobs," the colony's "most famous expatriate," a role model for future mariners, enabling Newfoundlanders to "share in the romance" of the American deep-sea fisheries.

Out of twenty skippers with obituaries in a section of Roberta Sheedy's Out of Gloucester website called "The Prosperity Makers," all are from either Nova Scotia or Newfoundland except for three Americans, one man from the Azores, and one from Sweden. Captain Ben Pine was born in Belleoram and some said he "appeared to have a natural talent for ship handling by virtue of just being born a Newfoundlander." Marty Welch came from Digby, Clayton Morrissey ("long-sparred Clayt," 193 centimetres tall) came from

Lower East Pubnico, Yarmouth County, and veteran master mariner William H. Thomas was born in Arichat, Cape Breton. As historian Michael Wayne Santos remarked in *Caught in Irons*, with upwards of a hundred years of family experience in the fisheries, these men brought skills that could easily be transferred to the American fishing industry. Most of them stayed in Gloucester and prospered. Lemuel R. Firth of Jordan Ferry, Nova Scotia, who became master of the *North Star* out of Gloucester at the age of twenty-two, went on to become director of the Cape Ann Bank and Trust Company, president of the Fishing Masters' Producers Association, and director of the Gloucester Fishermen's Institute. It was Captain Leo Hynes from Fortune Bay, Newfoundland, who had the honour of being the last dory trawling skipper to sail his schooner, the *Adventure*, out of an American port in 1953.

One Nova Scotian, who was not a captain, became most famous of all. Born in Port Medway, Queens County, in 1858, Howard Blackburn went to Gloucester to fish at the age of eighteen. In January 1883 Blackburn and his dorymate, Thomas Welch, a Newfoundlander, went astray in a blizzard from their schooner the *Gracie L. Fears*, and were stranded far from land on the Grand Banks. Left alone in the dory when Welch perished, Blackburn continued to row, his gloveless hands frozen to the oars. On the fifth day, after rowing over a hundred kilometres, he made it to Little River on the Newfoundland coast. He lost all of his fingers, half of each thumb, five toes, and the heel of his right foot. After many adventures, including sailing a sloop single-handedly to England and running a legendary saloon in Gloucester, he died in bed, aged seventy-three. Joseph Garland called Howard Blackburn "the greatest Gloucesterman."

> The wood of the vessel that will beat the *Bluenose* is still growing.
>
> —Angus Walters

Given their shared experiences on the deep-sea fishing grounds, Gloucester and Lunenburg, fishing capitals of their respective nations, naturally had much in common. They shared a history of the fishery as well as a great pride in themselves and in their vessels. But they also had their differences, rivalries that were usually friendly but not always so. That rivalry emerged most dramatically in the international fishermen's races held between 1920 and 1938.

Plenty of races had taken place between fishing masters and their schooners before the 1920s, mostly informal and spontaneous "hookups" when vessels met at sea or to and from the banks. Such "unofficial brushes" tested seamanship and made for great storytelling, said Michael Wayne Santos, "good yarning during mug-ups in the forecastle." An early race that attained "legendary stature" was held in 1892 during the celebrations for Gloucester's 250th anniversary of its incorporation. Because it took place in a howling August gale, it came to be known as "the race it blew." A "grizzled veteran of the fishery" declared: "that'r first Annivers'y Race was the only real man-eat-man race us fishermen have ever pulled off."

The competition that began in 1920 came about because of a race in which the wind blew hardly at all. To the disgust and amusement of fishermen everywhere, remarked Keith McLaren in *A Race for Real Sailors*, in July that year the New York Yacht Club postponed a race in the America's Cup series because of a twenty-three-knot wind. Why, it was said, "twenty knots of wind would barely press the wrinkles out of the sails of a banks schooner." The "snort of derision could be heard from Newfoundland to Gloucester!" wrote Brian and Phil Backman in *Bluenose*. "After that, the demand for a series of races between honest-to-god sail carriers gathered like an August gale." In *Witch in the Wind*, Marq de Villiers claimed that the real originator of the idea of the international races was Colin McKay of Shelburne; as a good socialist, he had poured scorn on the America's Cup "and its millionaires' frolic of fooling around with costly devices." Endorsed by the *Canadian Fisherman*, his suggestion was picked up by the Halifax press. William Dennis, owner of the *Halifax Herald*, obliged by raising four thousand dollars in prize money and putting up a silver cup to be awarded to the fastest schooner in the Nova Scotia and New England fishing fleets.

The first races were held off Halifax in 1920, Gloucester having enthusiastically accepted the challenge. To the chagrin, and perhaps surprise, of the Nova Scotians, the Gloucester contender *Esperanto* defeated the Lunenburg schooner *Delawana* under Captain Tommy Himmelman. The *Esperanto*, Captain Marty Welch, arrived home to the greatest cheering crowd that had ever gathered on the Gloucester waterfront and a whirl of banquets and receptions. Governor Calvin Coolidge, soon to be the thirtieth president of the United States (and who, on hearing of his death some years later Dorothy Parker famously remarked: "How do they know?"), declared the *Esperanto*'s victory to be "a Triumph for Americanism," conveniently forgetting Captain Welch's Digby origins, as Marq de Villiers has noted.

Something had to be done. Largely at the instigation of a group of Halifax businessmen, the Bluenose Schooner Company was formed, naval architect William J. Roué engaged as designer, and a fine new schooner, the *Bluenose*, was launched from the Smith & Rhuland shipyard in Lunenburg on March 26, 1921. Captain Angus Walters, who had agreed to be skipper as well as being the major shareholder, according to the Backmans declared her to be a "wery good wessel." The trophy would be returned to its rightful place.

And so it was. In 1921 Gloucester's *Elsie*, smaller than *Bluenose* and eleven years old, arrived in Halifax to face a schooner built at least partly for speed. In the first race "*Bluenose* streaked for home... like a kerosened cat through Hades," with her lee rail buried so deep "[w]e reckoned you could drown a man in her lee scuppers,"[9] enthused the press aboard the *Lady Laurier*. For the first time the international trophy was in Canadian hands. There was of course much jubilation in Halifax. The crew and supporters gathered in the forecastle and consumed some celebratory champagne, apparently delivered to the *Bluenose* by an American admirer. Eventually, the Backmans related, a growl was heard from one of the lower bunks: "Vell! Now det ve finished det t'underin' apple chuice, let's bwreak oudt de Wrrum and get down tuh sehwerious drrinkin'!"

Captains Ben Pine and Angus Walters, 1923.

Over the years, the atmosphere surrounding the races became increasingly acrimonious. The squabbling began in earnest when the *Bluenose* defeated the *Henry Ford* under Captain Clayton Morrissey off Gloucester in 1922. Howard Chapelle, naval architect and historian, according to John Morris declared the race to be "a noxious affair in which the sail plan of the *Ford* was repeatedly reduced by the race committee, to the point where her mainsail would not stand properly....All those [races] in which the Canadian Captain Walters sailed were distinguished by a complete lack of sportsmanship and by much bickering." According to Joseph Garland, Chapelle thought Walters was "an aggressive, unsportsmanlike and abusive man, but," he had to admit, "a prime sailor."

In some ways the 1923 race off Halifax was the best of them all, for *Bluenose* and the Gloucester challenger *Columbia* were well matched, the former winning both races by a mere hair's breadth. Dana Story in *Hail Columbia!* described the second race as "being little short of an epic." Nevertheless, controversy raged as usual. In the first race, Captain Ben Pine, rival and friend of Angus, forced Walters either to wreck the *Bluenose* on the shoals near the shore or to collide with the *Columbia*. Angus chose the latter course, swallowing *Columbia*'s wind as he crossed her bow and actually towing his rival briefly as *Bluenose*'s main boom caught in *Columbia*'s shrouds. Both captains had committed infractions and no protest was made. *Bluenose* won the second race but Pine accused Walters of passing a buoy on the wrong side. The race committee agreed and awarded the race to *Columbia*. Walters's suggestion that the race be declared no contest was refused, and in disgust he sailed for

Bankers racing, from the deck of the *Bluenose*.

home. Or, as W. M. P. Dunne, biographer of vessel designer Thomas F. McManus, put it, "an already sulking Captain Walters tossed a tantrum…and went home to Lunenburg in a high state of dudgeon." J. B. Connolly said Angus was "overcome with cold feet." Captain Pine declined to run over the course alone and so the series was declared incomplete.

At Gloucester in 1931 the *Bluenose* faced and defeated a new and final challenger, the *Gertrude L. Thebaud*, despite badly stretched sails and a poorly repaired keel after having earlier gone aground near Argentia in Newfoundland. Seven years later, in 1938, though the *Bluenose* was now seventeen years old and hogged, sagging slightly from the bow and the stern, Angus Walters agreed to race the *Thebaud* one last time, declaring, according to McLaren, that he "would race in a puddle of brimstone for a counterfeit Chinese centime to settle the championship once and for all." It came down to one final race, each vessel having won twice. In the end, the aging *Bluenose* triumphed, keeping the trophy forever. It was "her finest hour," wrote Lunenburg historian Heather-Anne Getson, her average speed just over fourteen knots, "one of the fastest speeds ever recorded over a fixed course by a canvassed vessel." As they listened to radio broadcasts back home, the *Halifax Morning Chronicle* of October 26 reported, "Tears of joy trickled down the weather beaten faces of Lunenburg's grizzled sea veterans" as their "pride and idol" retained her title for all time. As if to "shout their congratulations across the waves," the sister schooners of Lunenburg's fishing fleet "let loose a deafening blast from their sirens and fog horns, church bells, car horns and even the bell of the old town clock added their joyous voice to the din." It had been the last chance to defeat "the black-hulled beauty from Lunenburg."

Well before the final triumph in 1938 the *Bluenose* and Angus Walters had become celebrities, even legendary. A broadside view of the schooner in full racing rig graced a Canadian stamp in 1929. (Angus Walters would later have a stamp of his own, as would designer William Roué.) A schooner appeared on the Canadian ten-cent coin in 1937, and in 2002, as everyone had been assuming for years, the government declared it to be a depiction of the *Bluenose*.

In 1933 skipper and vessel sailed to Chicago for the World's Fair, the "Century of Progress Exhibition." Two years later the *Bluenose* sailed to England for George V's Silver Jubilee, where, according to Heather-Anne Getson, the British press labelled Walters "Captain of the Queen of the North Atlantic." Angus chatted with the king and reported that he was "a very nice, ordinary sort of fella." On the way home, on September 16, 1935, the *Bluenose* encountered a fierce hurricane a couple hundred kilometres off the English coast. At the height of the storm the vessel heaved over onto her beam ends, but after a pause, and with a shudder, she righted herself. Five women passengers not allowed on deck had just started to play a gramophone record of Cole Porter's "Anything Goes." The vessel had to limp back to Plymouth for repairs. Angus later recalled: "It was the biggest sea I ever saw, worse than the night off Sable when we nearly lost ourselves and the vessel. And for the first time in her life the *Bluenose* keeled over."

It's fair to say that the Americans held less exalted views of the *Bluenose* and her skipper. Big and powerful, they were not used to being outwitted by a wily little "Novie." Dunne referred to "that old curmudgeon, Angus Walters, *Bluenose*'s hardbitten master," calling him "a cantankerous Canadian-Irishman" (a change from being called a Dutchman). Connolly in *The Port of Gloucester* maintained that the *Bluenose* was "never a real seagoing fisherman"; despite considerable evidence to the contrary, he claimed that she "proved a flop as a fisherman." The grapes south of the border could be decidedly acidic. But as for her career as a racer, of the twenty-one official races she sailed in, according to Dana Story, she defeated fifteen different schooners and lost only six times. She was indubitably a winner.

In *Gloucester on the Wind* Joseph Garland spoke of the "bittersweet curtain call" of the international fishermen's races of the 1920s and 1930s. They represented in many ways the end of an era, a hard and colourful way of life that was on its way out. They had "gripped the imagination…the sea call in men, the nostalgia, the sentimental."

Sadly, the fishing schooners that had sought honour and fame in the international races, especially the American ones, too often came to tragic ends. Some maintained they were jinxed. In May 1921 the first champion *Esperanto*, coming home from the banks with a full load of salt fish on her first trip after her victory, ended up on the shoals of Sable Island in a thick fog; the crew was rescued by the *Elsie*. In her turn, *Elsie* sprang a leak and sank off St. Pierre in 1935, leaving her crew to row seventy-five kilometres to shore. The *Mayflower*, disqualified from challenging the *Bluenose* in 1921, tossed her skipper Captain Alvaro Quadros of Gloucester overboard to his death on the Western Bank during the August gale of 1927. The *Puritan*, built to rival *Bluenose* in 1922, ended up on Sable's Northwest Bar on her second trip in June of that year. In June 1928 a west wind drove the *Henry Ford* on to the Whale's Back Reef off Martin's Point, Newfoundland, and she was abandoned as a total loss. The *Gertrude L. Thebaud* was beaten to pieces on the breakwater at La Guaira, Venezuela, during a terrific storm in February 1948, just a few years after the *Bluenose* met her own fate off Haiti. This would appear to be a disastrous record, but these more widely known vessels were simply sharing the fate of a great many fishing schooners at the time.

The August gale of 1927 struck Sable Island and the surrounding waters on the twenty-fourth of the month and the coast of Newfoundland a day later. By the end of August the people of Newfoundland knew the extent of the damage to the small schooners fishing off their south coast. It was two weeks after the storm, however, before they knew that even more Newfoundland men had been lost on Nova Scotia vessels, particularly the *Joyce M. Smith*.

News came in slowly about the other vessels that had foundered, but by September 21, when the wreckage of the *Uda R. Corkum* was located, almost a month after the great gale, it was evident that four Lunenburg schooners had succumbed to the storm off Sable Island, with the loss of over eighty men.

There was, however, one more fishing vessel still missing. The handlining schooner *Columbia* had sailed from Gloucester under Captain Lewis Wharton on July 3 on her second trip of the season and put into the captain's hometown of Liverpool, Nova Scotia, on July 15 for supplies and additional crew. On the day before the storm, August 23, another Gloucester schooner, the *Herbert Parker*, Captain John Carrancho, had encountered the *Columbia* north of Sable Island on the edge of Western Bank and had been told that few fish had been caught there. The *Columbia* had not been heard from since, and in the second week of September five dories and a pair of grey oars marked "Columbia" washed ashore on Sable Island. Down in Gloucester, *Columbia*'s owner Captain Ben Pine said that sixty or so of the vessel's old oars had been sold to different fishermen and the oars might well belong to them; the schooner had been fitted with new canvas and had lots of supplies on board, and in any case she wasn't expected home before the beginning of October. But then, according to John Morris, more debris washed ashore on Sable, including Gold Medal flour, New Jersey brand milk, and oak hatch covers, all materials that would not be on a Canadian vessel.

On September 19, to the great relief of the people of the town, the *Gloucester Daily News* with banner headlines declared: "Columbia Safe after Storm." Captain Lemuel Spinney of the *Oretha F. Spinney* had sighted the *Columbia* a few days earlier; he'd been fishing for halibut on the Grand Banks and knew nothing of the storm, but had passed *Columbia* on Quero Bank on the way home. Captain Pine sent a reassuring message to Mrs. Wharton in Liverpool. But on September 27 the *Halifax Morning Chronicle* observed that Captain Spinney had not actually spoken to *Columbia*, since at the time he didn't know the vessel was considered missing, and, the paper went on to say, efforts to obtain information from Gloucester had resulted in little more than "a very ominous silence." On October 1 the *Halifax Herald* claimed that *Columbia*'s owners were still denying they were fearful for the schooner's safety. "I'm not worried," Ben Pine had said. Captain Wharton "is an able and masterful skipper, [and] knows the Canadian coast like a school boy knows his A,B,C.'s, having handled some of the best Canadian and American schooners afloat."

The *Gloucester Times* was less sanguine, however, noting that the coast guard cutter *Tampa* had been ordered to Sable Island to search for "the racer." On October 1 the paper's headline read "Fear Fate of Schooner Columbia," and went on to catalogue the many signs and omens, which it had been reluctant to publish previously. Almost a month later, on October 26, Lunenburg's *Progress-Enterprise* wondered, if the *Columbia* was lost ("and she surely is"), had she been subjected to the "Race Cup Jinx" that had brought down so many of Gloucester's contenders for the international fishermen's races? That same day the Gloucester haddocker *Mary Sears* landed at the Boston Fish Pier with a whole dory bearing the name "Columbia." Even more convincing, the dory contained a bait knife with the letter "M" carved on the handle, a knife known to have belonged to a crew member. The state of the dory's bottom suggested it had been in the water for some time. "No Hope Now for Columbia," concluded the *Gloucester Times* on October 29. No one would ever know what happened to her. Finally, more than two months after the storm, Ben Pine was forced to believe that his beloved *Columbia* had been lost, with all hands.[10]

The hands, as it happened, were almost all Nova Scotians. The *Halifax Herald* of October 29 reported that the crew of the *Columbia* was composed of "stalwart Canadians, who have braced the sea since school days." Captain Lewis Wharton was one of the most able skippers that Liverpool had sent forth, and "none knew the

THE GLOUCESTER CONNECTION

treacherous shifting bars of Sable Island better"; his wife and "two charming daughters" knew that "no braver man ever sailed out of Gloucester than he."

Most of the rest of the crew came from nearby Shelburne County, the largest number from the tiny village of Green Harbour, a community that had already lost several men that same day on the Lunenburg schooner *Clayton W. Walters*. Of the men who perished on the *Columbia*, Lyle Huskins had been so afraid the vessel would go without him that he had slept aboard her the night before she set out. Frank Dedrick, from Churchover, said to be one of the best violin players in the region, had been a fisherman from the age of fourteen. Enos Belong had planned for this fishing trip to be the last before he retired; his daughter Letitia lost not only her father but also her husband of less than a year, Clayton Johnson.

Captain Lewis Wharton, Liverpool, Queens County.

A particularly poignant story about the *Columbia*'s loss concerns the Firth family of Shelburne. Arthur Firth, head of a large family, had fished since he was a lad. Allister, his seventeen-year-old son, pleaded with his father to be allowed to stay home, saying he wanted to spend time with his chums before going back to school. In the end he told his father he wasn't going on the trip and put his gear on the wharf, but his father put it back. Allister then went to the home of friends and hid in a closet, but his father found him and said, "Boy, you are to come with me." Thus both father and son died

Crew of the *Columbia*, from Shelburne County. Back row (l to r): Allister Firth, Enos Belong. Front row (l to r): Clayton Johnson, Arthur Firth, Frank Dedrick, Lewis Wharton.

on the *Columbia*. Allister had been a member of the Juvenile Choir of Shelburne United Church and one of its leading singers. He was apparently "a bright boy and liked by everyone." His mother Bessie remarried and lived to the age of eighty-four. Arthur's uncle, Samuel Firth, who at one time had fished on the Grand Banks out of Gloucester, had gone down on the *Sadie A. Knickle* in the August gale of 1926.

As was usual on fishing schooners at the time, there were plenty of family connections. Joseph Mayo died along with his eldest son Albert, or Ab, and his youngest son George. Ab Mayo was in fact a last-minute replacement since his brother Joe wanted to stay home with his pregnant wife. A fourth brother, Bert, survived the gale aboard an engine-powered fishing trawler. About 1:00 A.M. during the storm Bert was startled to see an apparition of his brother George in the wheelhouse of the trawler. George looked at him a moment and said "Bert, we're all gone," then vanished. Although resident in Halifax at the time, the Mayos were from Burin in Newfoundland. Given the nature of the North Atlantic fishing triangle, it's not surprising that there were other Newfoundland connections. Captain Wharton had met and married a young woman from Harbour Breton, likely when his vessel came into port for bait. Rupert Bragg from Dorchester, Massachusetts, who was apparently the cook aboard the *Columbia*, was originally a Newfoundlander from Channel-Port aux Basques.

Several men had the good fortune not to sail on the fateful trip. Enos Belong had his nineteen-year-old son Jeremiah on board with him, but the younger Belong, fortunately for him, developed an abscess on his neck and had to be landed ashore. A man named Basil Williams came down with blood poisoning and had the same good fortune. Robert Nauss of Gloucester, a student at the University of Pennsylvania, had his clothing on board ready to sail when Captain Pine reminded him that the *Columbia* would not be back in time for him to return to college; Nauss tossed his clothes on the wharf and leapt ashore at Gloucester just as the vessel was pulling away from the pier.

Albert Mayo, Halifax.

With the loss of twelve men on the *Columbia*, the toll of dead from Shelburne County that year came to twenty-three, including the ten men on the *Clayton W. Walters* and one on the *Julie Opp*. In October, four more Shelburne County men died on the *Avalon*. The *Shelburne Gazette and Coast Guard* on November 3 called 1927 "A Season of Gloom."

> ## Lost on the *Columbia*, August 24, 1927
>
> **FROM LIVERPOOL, QUEEN'S COUNTY:**
> Captain Lewis O. Wharton, 58
> James McLeod, 65
> George Williams
>
> **FROM HALIFAX:**
> Joseph Mayo, 54
> George Henry Mayo, 28, son of Joseph Mayo
> Albert Mayo Jr., son of Joseph Mayo
>
> **FROM SHELBURNE COUNTY:**
> Enos Locke Belong, 54, West Green Harbour
> Samuel Belong, 37, West Green Harbour
> Frank Edgar Dedrick, 54, Churchover
> Arthur Firth, 62, Shelburne
> Allister Cecil Firth, 17, flunkey, son of Arthur Firth, Shelburne
> Thomas Richard Hayden, 39, Shelburne
> Lyle Huskins, 20, West Green Harbour
> Clayton Monroe Johnson, 26, West Green Harbour, son-in-law of Enos Belong, brother-in-law of Carrol Williams
>
> Foster McKay, 20, West Green Harbour
> Robert Burns Stuart, 55, West Green Harbour
> Carrol Lyle Williams, 28, West Green Harbour, brother-in-law of Clayton Johnson
> Nathan A. Williams, West Green Harbour
>
> **FROM PARRSBORO, CUMBERLAND COUNTY:**
> James McAloney, 24
>
> **FROM GLOUCESTER:**
> Isaac Gould, 60
> Colin Hawley, 30
>
> **FROM BUCKSPORT, MAINE:**
> William Colp
> Leo White
>
> **FROM DORCHESTER, MASSACHUSETTS:**
> Rupert Bragg, 46, cook

The *Columbia* was a great loss. Among other things she was a fisherman, the last of the salt bankers to sail out of Gloucester without power, and therefore the last of her kind. She was also a racer, and expected by the Americans to take on the *Bluenose* and bring home the silver trophy. Joseph Garland said the *Columbia* was

THE GLOUCESTER CONNECTION

"the best we put forward." Angus Walters, and others, considered her to be the finest challenger of them all.

But above all, perhaps, she was beautiful. As Dana Story, son of Arthur D. Story who built her in Essex in 1923, put it: "a stinking salt fisherman one day, she could be transformed into a white-winged creature of beauty the next." Gordon W. Thomas in *Fast and Able* described her as "fast, famous, and ill-fated…the last vessel to sail salt fishing from Gloucester; and she carried more men at one time to a watery grave than any other vessel out of this port." She was "the Gem of the Ocean," he concluded, "or perhaps we should say, the Flying Dutchman, ship of disaster." Silver Donald Cameron in *Schooner* described her as "one of those ships so sweet in her

Columbia

lines, so well favoured in her bearing, as to seem slightly unreal, like something beyond the capability of mere men to create." And James Connolly declared her to be the most beautiful schooner he ever saw. "And to see her under sail and coming bow on in a smooth sea and a fresh breeze—to see her so, viewing her from under her lee bow, and the way she had of easing that bow in and out of the sea—well, the beautiful lady was Poetry herself then."[11] *Columbia* had existed just fifty-two months.

Captain Ben Pine late in life said that the *Columbia* was "the finest piece of wood ever to show its 'highs' out of Gloucester." He also declared, when she was lost, that "Never again will I send another craft fishing unless she has an engine." "Piney"—"easygoing and humorous," said to be a real gentleman—died in 1953 and is buried, according to Dana Story, in the Oak Grove Cemetery in Gloucester. Above his name on the monument, "under full press of sail and bending to a breeze of wind," is a representation of his favourite schooner, the *Columbia*.

On the morning of January 3, 1928, the National Fish Company steam trawler *Venosta*, Captain Gjert Myhre, returned home to Halifax, her trip cut short by the loss of about six hundred dollars worth of trawling gear. Fishing a hundred kilometres or so west-southwest of Sable Island shortly before 2 A.M. on January 1, she had been steaming along dragging her nets on the bottom in about forty fathoms of water when suddenly she was stopped dead in her tracks. Slowly her great steam winches began to wind in the taut, heavy cables. The men on the *Venosta*, speechless at first, rushed to the rails as two masts, like index fingers, broke through the crest of a towering swell. Shortly the hull itself, on a perfectly even keel, emerged above the waves. "No more ghostly dramatic story has for a long time come from the great deep," declared the *Lunenburg Progress-Enterprise* on January 4, "than that which described how in the faint misty moonlight over a waste of waters, a noble lined vessel with her stately masts intact, her shrouds complete, but minus sails and booms, rose out of the calm depths of ocean to a turbulent sea." No one had to be told she was the *Columbia*, for she was "different from any other fisherman in the fleets."

It was an eerie sight. The moon had set, Dana Story related, but the sky had been swept clear by the gale. Three or four vessels could be seen on the horizon, and to the southward were the blazing lights of a liner. The floodlights of the trawler outlined the deck of the spectre, and a hush fell over the crew as they saw the blackness of the open companionway leading into the forecastle, where bodies must once have lain. The steel tentacles of the trawler held her and the two vessels plunged and crashed together in the rising seas, imperilling both. Poised briefly amid the swirl of whitened waters, the schooner "surveyed again the elements that had destroyed her, appeared as a sentient thing, trembling with resurrected life, had the wind shake again her shrouds, her decks washed again by the merciless waves." Then, "as if in weariness of spirit," she "resignedly sank back to her deep silent grave, the waters madly roaring down the black opening of the forecastle companionway, a ghostly sight in the wan light of the moon." Amid crashing gear and the snapping of giant cables the phantom ship slipped back into the depths that would hide her forever.

the CALL of the SEA

> The horizon's curve had caught, and held his eye
>   With fascinating gleam…Far away
>   The Atlantic thralled his soul with mystery,
> Beckoned him on, and called him by his name…
>   The stories of the Banks, the travel lore…
>   Had cast their spell upon him.
> —E. J. Pratt, *Rachel* (1917)

THE ECCENTRIC AMERICAN WRITER GERTRUDE STEIN, living in Paris in the 1920s, may or may not have spoken the words "lost generation" to describe the young people who had served in the First World War. At least Ernest Hemingway credited Stein when he used them as an epigraph in his 1926 novel *The Sun Also Rises*. Later, however, after a notable literary feud, Hemingway maintained that the term had originated with the garage owner servicing Stein's car, who had referred to his rather feckless and poorly skilled young mechanics as *une génération perdue*. The phrase came to have various meanings, but in general it referred to the men who came of age during the war, the young men in their twenties and thirties who had fought in the war and who had emerged disillusioned and cynical about the world, uncertain of their place in it. In Britain the phrase was originally used for those, particularly of the upper classes, who had died in the war, the "flower of youth" gone forever, the "best of the nation" destroyed, robbing the country of a future elite. George Orwell put it more bluntly: the war "carefully selected the million best men in England and slaughtered them, largely before they had time to breed."[1]

The words "lost generation," therefore, can refer to the men who survived the Great War and were deeply affected by it or they can be used to describe those who were doomed to die in the war, leaving behind a huge gap in their society. Many of the North Atlantic fishermen who died in the August gales of the 1920s had been touched by the war in one way or another. Men of their generation had fought in the war and too many had not come back. Some of the Nova Scotia fishermen caught up in the war had volunteered to fight, but more had faced the trauma of conscription under Canada's *Military Service Act* of 1917. Fishermen, like farmers, had been considered essential for the production of food and therefore exempt from military service, but in April 1918 the exemptions were removed by an order-in-council. By that stage of the war the conscripts knew they would likely end up as mere cannon fodder and it's not surprising that some of them resisted. Though many Newfoundland soldiers died in the war, there was no conscription there; the Americans did not enter the war until the crucial final year.

But it's the other interpretation of the phrase "lost generation" that perhaps best defines the fate of the fishermen who were lost at sea in the 1920s. Admittedly on a microscopic scale compared to the millions who died in the war, the loss of a generation of mainly healthy and enterprising younger men nevertheless left gaping holes in many of the small communities they came from. The owners of the vessels obviously suffered financial losses, but it was the small coastal villages that supplied the schooners' crews that suffered the most: Blue Rocks and Stonehurst, Indian Point, Vogler's Cove, the LaHave Islands and other communities along the LaHave River in Lunenburg County, West Green Harbour in Shelburne County, the outports of Fortune and Placentia Bays in Newfoundland. Wives had to go on living without their husbands and children would grow up without their fathers. A generation of children would not be born. The tragedy was made worse by the fact that family members often sailed together, thus the storms took not only fathers and husbands, but sons and brothers, brothers-in-law and cousins, as well as friends and lovers. Blue Rocks, for instance, was described by Errol Knickle, a later resident, as "a village of widows," a whole community of women left to cope on their own.

As in the war, it might be said that the men lost were "the finest and best in the county," as the *Lunenburg Progress-Enterprise* had put it. The men who went to sea were usually breadwinners but also men with ambition, the adventurous ones. In 1926 Warden Romkey of Lunenburg County had lamented that the storm had taken "the best and brightest of our youth…the ablest, finest men of the community." And the *Halifax Herald* of October 10, 1927, echoed the sentiments, noting that "the angry elements of the sea" had snatched from this place over eighty men, "swallowed up so many of the flower of our fleet."

The number of fishermen lost in the August gales of the 1920s surpassed 200. Fifty men died on the two Lunenburg schooners that went down in 1926, at least 84 more on the four that foundered in 1927; the total number of men lost on Lunenburg vessels was therefore 134 (the figure may be slightly higher because it's not known for certain just who was on board these vessels). More than 40 Newfoundland fishermen perished when the gale struck the south coast of the island in 1927; when one includes the 23 Newfoundland men who died on Lunenburg's *Joyce M. Smith* and the *Uda R. Corkum*, the figure for men lost from the island that year is over 60. As for Nova Scotia fishermen, 50 died in 1926 and 61 more in 1927, plus another 19 (along with 5 New Englanders) in 1927 on Gloucester's *Columbia*, making for a total of at least 130 Nova Scotians lost in two years of fishing. Given Captain Quadros's death on the *Mayflower* and the certainty that some men were lost but not counted, especially in Newfoundland, the figure is easily over two hundred. The August gales had certainly taken their toll.

After the tragedies there was some attempt to compensate the families for their losses. In Nova Scotia most of the dependants were eligible to receive funds from the Workmen's Compensation Board. According to historian Fred Winsor in his article "Solving a Problem," the government's response was miserly. Widows received $30 a month for life, or until they remarried, and $7.50 a month for children under sixteen. One woman in the LaHave Islands lost two unmarried sons in the 1927 gale; for the son who was an experienced fisherman she received $10 a month for several years, and $5 for the other son. Women whose husbands were lost—required to remain single or lose the compensation—had to face a long solitary life or live unmarried with a new partner. Some women who lost their husbands found it necessary to leave their homes and move in with other widows and their families, the combined funds enabling them to survive. Nita, the wife of Hastings Himmelman, having been widowed at the age of twenty-eight when her husband died on the *Sylvia Mosher* in 1926, was receiving $50 a month when she died in 1961. The disastrous fleet losses had nearly bankrupted the provincial fund. Instead of taking steps to improve safety, historian David Frank wrote, the Nova Scotia government was concerned about the high cost of compensation claims: "[A]s a result fishermen were written out of the existing Workmen's Compensation Act and turned over to private insurance companies."[2]

The twenty Newfoundland fishermen lost on the *Joyce M. Smith*, as well as the three men who died on the *Uda R. Corkum*, were not eligible for compensation in Nova Scotia unless they were residents in the province. Fortunately a Permanent Marine Disaster Fund had existed since earlier in the century. The *St. John's Daily News* led an appeal for cash to be added to the fund in light of these losses and the men who had died off the south coast. Its editorial of September 1, 1927, declared that "the widows, the aged mothers and the fatherless little ones are suffering.... Theirs is a sickness of heart that even time cannot heal; a cold which the warm throb of practical sympathy may lessen; relief that, whilst it will not bring luxury and a superfluity of life's comforts, will mean all the difference between carking despair and new-born hope." Contributions flowed in: $500 each from Bowring Brothers and the Bank of Montreal, for example, $250 from His Grace E. P. Roche, archbishop of St. John's. The Girl Guides of Grand Falls collected $2.49. Dr. John W. Niven, an English Presbyterian minister and elocutionist, proud kinsman of Robert Burns, en route to the United States for a lecture tour, gave a special Shakespearean recital despite a nasty cold to help the fund. Manoel Joaquim de Carvalho & Cie., of Bahia, Brazil, hearing of the terrific gale, sent £100. The fund was able to give $40 a year for five years to young widows, the same amount to aged widows and in some cases to aged parents for life, while $25 yearly was paid to each child until his or her fifteenth birthday.

Meanwhile, the people of Gloucester were exhibiting similar generosity. Once the *Columbia* was confirmed lost, owner Captain Ben Pine began to receive letters from Nova Scotia. Mrs. Arthur Firth of Shelburne wrote: "As my husband was in your vessel Columbia and I am left destitute with two small children and no means whatever for support. The men out of Lunenburg that was lost in Lunenburg vessels draws a compensation fund but lost souls understand there is none for us. So have been advice to write you to see if you could not give us some support....I am a widow along in years and not able to work much. The ages of my children are 14 and 15. As you know, I had my husband and son both lost in your vessel. I will expect to hear from you soon."

Frank Dedrick's widow wrote as well: "I am left alone with six small children, left destitute, no home, and homeless. As my husband was lost in your vessel the Columbia I think it is about time something was done to help us poor souls out...." When news came in October of yet another loss, the *Avalon*, the American international race committee sprang into action, hoping to raise twenty thousand dollars for the relief of the families of the men who had "paid the supreme penalty in the hazardous employment of their calling." Both Captains Pine and Clayton Morrissey gave one hundred dollars to give the fund a boost. Businesses, citizens, fraternal organizations, fishing firms, captains, and the fishermen themselves gave generously, and within six weeks the appeal was oversubscribed by three thousand dollars. The Atlantic Supply Company of Gloucester even sent a representative to Shelburne and Liverpool to take a survey of all the families affected by the disaster. At Christmas, money and baskets flowed north, and soon notes came back thanking the race committee for the money, the food and clothing, and the payment of old debts. Said John N. Morris in *Alone at Sea*, "Gloucester had shown its heart to the Nova Scotia fishermen."

Much has been made of the romantic nature of the deep-sea fishery, the ruggedness and adventurousness of the fishermen, the courageous deeds and the comradeship, and, in the words of Sterling Hayden, the "soaring, mind-boggling beauty of the schooners." Wooden ships and iron men! Frederick William Wallace, in *Roving Fisherman* in 1911, was caught up in the wonder of it all: "These fishermen were seamen, colorful and robust. These handsome, able schooners were the survivors of the clipper ships of old. Here was salt water, wind, rope, canvas and hardy men—the main ingredients of the sea's romance. At least, I fancied so at the time."

Writing about the Grand Banks schooners in 1969 in *They Live by the Wind: The Lore and Romance of the Last Sailing Workboats* (and graciously including the fishermen of Lunenburg as well as those of Gloucester), American journalist Wendell P. Bradley spoke of "a breed of seamen whose hardihood and skill have never been surpassed." And the captains, he claimed, were "demigods." With their reputations earned as sail carriers or as great fish killers they were "ranchers of fish, students of wind and sea, intimate companions of waves, lordly, independent, sovereign—each voyage an accomplishment, each encounter with storm a victory, each passage home a contest. Their occupation gave a gloss to life, a glamour, an allure, an aura of excitement and suspense. It made of life a series of scenes, a performance."

In its review of the year in December 1920, the *Halifax Herald* described the extensive Maritimes seacoast as a "magnificent fishing field" in which the Nova Scotian was "lord paramount"—"hardy stalwart men" who braved "the utmost perils of the deep." "It has been said," the newspaper continued, "that the Nova Scotian is first and foremost a fisherman. Certainly, while every Nova Scotian is deservedly proud of our resources of farm and mine and forest none of these industries is fraught to him with the sense of romance and adventure that lingers about the fisheries. The mystery of the sea clings to the industry that lives upon its waters."

In *Young Men and the Sea* Daniel Vickers observed that in a world where "mechanism, routine, and calculation" were dominating life more and more, the sea seemed to be a place still ruled by forces that humans couldn't pretend to control; it was a place "where friendship and heroism, as well as evil and terror, still governed human life, and where the sublime ruled over the mundane. During the middle decades of the nineteenth century, the sea acquired a romantic aura… that it has essentially never shed."

Many have claimed, with some justification, that the whole idea was bosh, that there was nothing at all romantic about life at sea, certainly not in the North Atlantic banks fishery. "Romance of the sea and distant shores?" queried Joseph Garland in *Down to the Sea*. "None, unless you didn't have to go fishing for a living." In *Unto the Sea* Newfoundlander Garfield Fizzard reflected on the call of the sea: "The schooner fishery has often been described in romantic terms, but there was little romantic about the lives of bank fishermen. True, there was nothing on the ocean more beautiful than a schooner under full sail. When the sun rose warmly over a calm ocean, revealing schooners and dories scattered on all sides as far as the eye could see, there was beauty and tranquillity that lifted the spirit and moved men not given to lettered poetry to say, 'Ah, b'y, the water gets in yer blood'....More often, wet and intensely cold, they worked, frequently to the point of exhaustion."

As Michael Wayne Santos remarked in *Caught in Irons*, fishermen were not "romanticized caricatures," but "men very similar to workers in less exotic settings." Stripped of its romanticism, he maintained, "the pitching deck of a schooner was a place of work, in many ways no different from the coal face in a mine or the shop floor of a factory." "Working men who got wet" was the apt phrase used by Rosemary Ommer and Gerald Panting as the title of their book about sailors in the merchant marine; it applies equally well to the lot of the deep-sea fishermen.

But perhaps an old saying from Tancook Island in Mahone Bay, quoted in George Bellerose's *Facing the Sea*, says it best: "There is beauty and freedom in the fishing life, but there is rarely time to savour it. Fishermen, the saying goes, have only two guarantees: a tired back and a 'wet arse.'"

In the vast and stormy North Atlantic there were plenty of dangers beyond fierce tempests and shipwrecks. The Grand Banks, where warm and cold currents met, were especially prone to fog—the grey terror—one of the worst enemies of the fishermen. The men in the dories were most at risk, for it was all too easy to drift far from the schooner or to head off in the wrong direction. Lost at sea in a dory, going astray, meant exposure to the elements, a desperate need for food and drink, disorientation, and often death. Young Johnnie Angus, the protagonist in Arthur Hunt Chute's novel *The Crested Seas*, adrift on the banks with yet another squall on the way, reflected: "In that moment there came to me an appalling sense of the hate and might of the sea, such as could come only in a tiny dory. As we rode up and down upon the gigantic undulations above the fifty fathom bank we were mere chips at the mercy of the elements. Never before had a mere man seemed to me to be so puny, so ineffectual."

Several Newfoundland fishermen faced death at sea in July 1927. According to the *St. John's Evening Telegram* of July 8 there was "a southwest breeze with a choppy sea prevailing at the time." Two of the men on the *Marian Belle Wolfe* were drowned when their dory upset; the body of one of them, Martin Quann, was later found hooked in the trawls. In the same storm Charles Williams from Pool's Cove and George Robert May from Point Rosie near Garnish, Fortune Bay, both in their early fifties, were on the Lunenburg fishing schooner *Donald A. Creaser*, Captain Ellison Creaser. They had left the vessel to haul their trawls, were cut off by dense fog, and vanished, clearly having gone astray. Eleven days later, still adrift in their dory, having survived on twelve hard cakes and tobacco that had eventually run out, they were rescued by the steamship *Albuera* and taken to London. In Robert Parsons's *Survivors and Lost Heroes*, Williams recounted their ordeal: "It was just like hell...The preachers tell us about fire and brimstone and great thirst, sharp hunger and pain and agony. Well, my mate and I, we know something of what the preacher means. We didn't have the fire and brimstone, but there was sun that peeled our faces, seas that soaked us to the pores, icy winds that set us to shivering till our teeth clacked. After awhile, nothing to eat, and thirst—raging, tearing, maddening thirst—until we came across a whole ice-box, an iceberg. Yes, we know what hell is like."

A young man who went astray one February, nineteen-year-old George Reynolds on the schooner *Ida May* out of Gloucester, put both of his feet over the side of the dory into the water to keep them from freezing. He was eventually picked up by a steamer. A story of similar courage concerns a man from near Belleoram on the south coast of Newfoundland. Adrift and alone in the vast Atlantic, driven to desperation with thirst, he chopped his fingers with his splitting knife and sucked them. He too was eventually rescued, and continued to go fishing.

Poor visibility in thick fog led to many disasters. Vessels in stormy seas on crowded fishing grounds sometimes collided with each other. They were also in danger of being run down by faster and larger ships such as steamers and even ocean liners. The White Star Line's *Majestic* sliced through the Burin schooner *Antelope* on July 20, 1894, killing two of the crew. The Cunarder *Saxonia*, just out of Boston, ploughed into the *Mary P. Mesquita* in October 1899, cutting her in half. And in June 1925 the Gloucester schooner *Rex* was cut down on Quero Bank by the Cunard liner *Tuscania*. On a fair October morning in 1927, just as dawn was breaking, the Gloucester fishing schooner *Avalon* was run down off Cape Cod by the huge Italian Cosulich liner *Presidente Wilson*. The vessel sank in nine minutes, with most of the men trapped below deck. Although three men were rescued, "at least nine hardy Bluenose fishermen" succumbed to the "ruthless Atlantic."

A great many fishermen met their deaths by drowning. Vessels sank beneath them. Overloaded dories could be swamped in heavy seas. Men were washed or knocked overboard by flailing booms or slatting sails. An old fisherman once confided in a low voice to Ralph Getson at the fisheries museum in Lunenburg that a man on the trip who'd gone overboard had in fact been "making water"; no doubt other fishermen ended up in the drink while urinating over the side.

THE CALL OF THE SEA

On May 18, 1909, Mrs. Alexander Greek of Blue Rocks received the sad news from Captain Benjamin C. Smith, writing from the Magdalen Islands, that her son had been lost on the Lunenburg schooner *Gladys B. Smith*:

Dear Mrs Creak

I can heardly write thoes few lines For I feal so sorry   I cant expres my fealing to you Of the sad Accodent that happened on the 11th of May The Drowning of your Son Frazer   I know when you hear of it that it will break your heart   I can almost put myself in your place, but Accodence will happen, It seems Sometimes that it is the will of God

   There was a boat coming to help them, But he did not hold on   That boat got Edward Knickle & he feals very Sorry about it But he said he could not help It, he tried heard to save Frazer But he could not

   I suppose that you think It was to rough to be out   None of us thought of any thing like that to happen that day   The hole crue feals aful sorry About your Son   He was liked by Everybody   They all have the greatest Simpaty for you & his Sisters & Brothers   I know it is a heard Burden to bere, you will hear more About it when we come home If God spears us

   I do hope that the Lord will help you For he is the only one can   Till you meet your Son in Heaven, Where all Sorrow is ended

yours Truly

B. C. Smith

Ralph Getson suspects that letters like this one were not all that common. The tone of the letter, he maintains, says a lot about the character of this captain, for the more usual attitude was that there were plenty more men on shore to fill the spots of men who had been lost.

Despite the high probability of drowning, few fishermen knew how to swim. In the bitterly cold water they knew they couldn't last for long, and their heavy clothing would drag them down. And most often there was nowhere to swim to. Knowing how to swim simply meant that it took longer to drown.

> When for a moment, like a drop of rain,
> He sinks into thy depths, with bubbling groan,
> Without a grave, unknelled, uncoffined, and unknown.[3]

Alan Villiers, in his book *Wild Ocean*, portrayed dory fishing from schooners as "one of the hardest ways to make a living at sea that could well be devised." The poor doryman, he said, had to set off in "his little box of a boat" early every morning for months at a time, taking his chances against squalls, ice, fog, collision with steamers, "and the usual hazards of being overturned, swamped, or otherwise drowned…they are men against the sea, if ever there were any." "Let them stick long enough to the fishing," wrote Connolly in *The Book of the Gloucester Fishermen*, "and the sea will get them anyway: but it is every man to his own ending; and there are worse graves than the clean, green sea."

Fishermen aboard the schooners faced not only death but injury, often caused by long hours of work and lack of sleep. Accidents were bound to happen, sooner or later, on a slippery deck that was always moving, with so much gear scattered about. Heavy equipment could fall from aloft, and men themselves fell from the rigging to the hard deck below. The crew's hands were particularly vulnerable, for it was not always possible to wear protective gloves. Nets had to be mended, trawls with sharp hooks baited and hauled. Cuts became infected. Joseph Garland told of fishermen "Downed by the dread erysipelas, St. Anthony's fire, a horrible gangrenous infection of the hands, arms and face from handling fish in unsanitary conditions."

Injured crew members had to wait until the vessel returned to port to be treated. No one on board knew much about medicine, nor were they trained to administer it. There was also little life-saving equipment aboard. Seasickness simply had to be endured. Matthew Mitchell, who fished out of Lunenburg in the 1930s and 1940s when conditions were presumably somewhat improved, revealed in *Skipper* that "If you got sick aboard, you had to be pretty sick for them to bring you in. You didn't have nothing aboard. Anything Epsom salts, iodine or Friar's Balsam couldn't cure, that was it."

Given the roughness of the life, a life not suited to all men, there could on occasion also be a sickness of the mind. A bizarre incident recorded in Adolphus Gaetz's diary, no doubt unusual but indicative of the stress fishermen could face, occurred on October 9, 1871, at Mount Pleasant, Lunenburg County. Solomon Wamback had just returned from a fishing voyage and upon arrival he "indulged very freely in strong drink," the effects of which "it is supposed had caused Delirium Tremens." With a recently sharpened axe, he murdered his wife and four small children while asleep in their beds. The baby's head was nearly severed from its body. Wamback then tied a twenty-pound stone around his neck and another fifty-pound stone to his feet and dropped into the well.

James Fenimore Cooper, in *The Red Rover*, remarked that "Superstition is a quality that seems indigenous to the ocean." Given their inability to control the forces of nature, it's not surprising that many fishermen were superstitious. According to Kenny Reinhardt,[4] a fisherman from the LaHave Islands, a fellow must not wear grey mittens aboard the vessel or mention the word "pig," for it could make the captain "some savage." Such recklessness could bring on a rainstorm or worse. Some captains, he said, did not like music aboard: "It made them right spiteful because it was their belief that whistling could bring a gale of wind." The old skippers, he claimed, would never turn their vessels backward against the sun, and leaving Lunenburg harbour they'd never turn to the left but to the right, "making three parts of a circle before they headed out to sea." If a skipper saw a crow on the way out of the harbour he'd change his mind and fish on another bank altogether. You could get a "stiff kick in the arse" if you turned the hatch covers upside down. One old vessel skipper would look over each plate of biscuits the cook brought in, and if one of the biscuits was upside down he'd turn it right side up. There were some fellows known as "Jonahs" because they seemed to bring bad luck; they were either landed or not hired in the first place, said Reinhardt, because "they'd get no fish."

There were plenty of other superstitions. Sailing on the thirteenth of the month would bring bad luck, as would thirteen letters in the name of a vessel, whereas many schooners' names included three of the same letter to bring good fortune. A schooner should not be launched on a Friday. A vessel might go bottom-up if a loaf of bread was turned upside down during the voyage. And hammering a nail on a vessel on Sunday could bring misfortune. A curious Lunenburg superstition was discovered by the folklorist Helen Creighton: if, for whatever reason, someone didn't want a ship to sail, they would put a black cat under a basket and keep it there.

Frederick William Wallace claimed that Newfoundland fishermen had a custom known as "crossing the seas" to avert the menace of a huge wave. When lying-to in a gale and seeing a big sea roll up towards the vessel, they would make the sign of the cross in front of the advancing comber and murmur, "May the Cross of Christ come between you and us." Grand Bank historian Robert Parsons confirmed that Newfoundland fishermen made the sign of the cross before threatening waves. Marq de Villiers suggested that captains generally thought prayer was for funerals and weddings, "for a real skipper trusted his men and the hull he stood on and the rig he controlled and his own very considerable skills." The *Bluenose*'s Angus Walters once declared, however: "There were times when a little religion was mighty handy. The fo'c'sle is a damned poor place for an atheist because he'd find little company." As Silver Donald Cameron alleged, "Superstition and religion are indeed parallel and logical responses to the marine experience."[5]

New England fishermen were equally superstitious. Procter Brothers's *Fishermen's Own Book* in 1882 claimed that a belief in lucky and unlucky sailing days was universal, Friday especially being an ill-omened day on which to begin a voyage. Jonahs could be animate or inanimate objects, a man or a vessel, indeed almost anything supposed to bring bad luck. Some believed that a valise, a most unseaman-like article for a fishing trip, when carried aboard was a Jonah, as were violins, checkerboards, toy boats, and a bucket sitting on deck partly full of water. Fishermen whistled for a breeze when it was calm, and stuck a knife in the after side of the mainmast to bring a fair wind. A bee or a small land bird coming on board brought good luck, whereas ill fortune followed the lighting of a hawk, an owl, or a crow on the rigging. A hook that had been stuck in the flesh of a hand was immediately thrust into a piece of pine to avoid soreness in the wound. Wesley George Pierce in *Going Fishing* maintained that some thought it was unlucky to turn a hatch-cover bottom-up or to drop it in the vessel's hold. It was unlucky as well to take an umbrella on board.

A ship is worse than a gaol. There is in a gaol, better air, better company, better conveniency of every kind; and a ship has the additional disadvantage of being in danger.

—Dr. Johnson to Boswell

It must be said that conditions for living on board the schooners could be less than ideal. There was, first of all, limited space, notably in the forecastle where the men ate and slept. As Fred Winsor observed, conditions aboard were primitive. The crew members had to share the cramped space with twenty or more other men for months at a time, with only one spare change of clothes for the duration of the voyage. All the fresh water had to be used for cooking or drinking, not for washing oneself or one's clothes. Collie Greek told about his father Howard coming home after months away on a trip and setting out in a dory from the harbour in Lunenburg for the trip back along the shore to Blue Rocks. His wife would see him coming and take clean clothes down to the fish store, making him strip and wash himself there. The clothes he'd been wearing all summer were put in a tidal pool, held down by rocks, and left there through several tides to get the smell out of them. Finally they'd be clean enough for his wife to wash.

Commenting on a trip he'd taken aboard a schooner, Frederick William Wallace remarked on "the rank odor of disturbed bilge-water mingled with tobacco reek, tarred lines, oil-skins and fish… the effluvia from disturbed bilge-water swishing amidst the slimy ballast under the floors, was strong enough to choke a skunk." In the words of Keith McLaren, in *A Race for Real Sailors*, "The smells of the stove, sodden wool clothing and as many as twenty working men—blending with the aromas of tobacco, fish, cooked food and the pungent bouquet of fishy brine from an unpumped bilge—made life in the forecastle an experience for the senses." As Errol Knickle of Blue Rocks once said, it would be hard to say what smelled worse, the bilge or the guy in the bunk above.

Then there was the little matter of dealing with the fact that the vessels had no head. A skipper turned guide at the fisheries museum in Lunenburg, when asked what one did in the circumstances, pointed to a pail and said "bucket and chuck it!" And in rough weather? Well, "you'd get a wash at the same time." Kenny Reinhardt said if it wasn't too stormy the business could be done in the hold; on deck in a storm, though, you could slide from one side of the deck to the other and back again. Another trick was to go behind the wheelhouse and "blame the dog"!

"Life in a banker then," wrote G. J. Gillespie about the 1890s in *Bluenose Skipper*, "was two-fisted. Only men with stamina—the kind of fortitude that defied physical discomfort and ever-present danger on the sea—could play this game that demanded so much of body and mind."

If it was all so bloody awful, why did men go to sea?

The simplest answer, perhaps, is that fishing was a way of life. Boys grew up around fishermen, learned from them, and listened to their stories about the sea; from early days they spent time around dories and watched the schooners leave the harbour and return with fish. By the time they were young men they knew about fishing gear, about winds and tides, and they knew something of the risks too. "The sea is before their eyes from infancy," wrote Wallace, "the roar of it is in their ears and the smell of it in their nostrils."[6] In his long poem *Rachel*, E. J. Pratt spoke of "impetuous cravings for the restless sea." It's also likely that young men had fathers or older brothers or uncles who had gone to sea, and eventually they would sail together. Sometimes fathers insisted their sons go fishing. Often young men went because their buddies did.

A reporter in the *New York Sun* in August 1928 observed that "Nova Scotia—New Scotland—is practically an island entirely surrounded by fish....It is not for nothing that a fat cod is the central ornament on the coat of arms of the old province." (In fact, the fish was a salmon.) Lunenburg was the very heart and centre of Grand Banks cod fishing, Sable Island the "burying ground of the fishing fleet." Like the men of Gloucester, the reporter declared, Lunenburgers knew the risks, and they knew the profits were small, but there was never any lack of recruits for the fishing fleet. "It is the ruling ambition of the boys and their first trip with the fleet is the grand event of their uneventful lives. It gives them a set and swagger that no other sort of men seems to have, in these parts, at least."

Obviously many men went fishing for economic reasons, needing money to support their families. In short, it was what they had to do. Ervin Langille of Tancook Island explained why he first went on the schooner *Florence B.* to the Grand Banks in 1913, age eighteen: "I had no gear to fish at home so I went....I had to do something for a living, and once I got married, I had two, three, then four children. After a while it came to nine that we raised."

Garfield Fizzard, writing about his uncle Captain Frank Thornhill, observed that history in Newfoundland is inextricably intertwined with the catching of fish. Captain Thornhill's career began as a doryman aboard a wind-powered schooner and ended as captain of a steel trawler; he was one of the last representatives of an extraordinary group, the masters of banking schooners. Fizzard claimed that the "ambition of every young Fortune Bay man was to be a bank fisherman. It was not that banking was easier or more lucrative than inshore fishing. It was more a rite of passage that drew them like a magnet, giving them a more independent position in the family." No longer simply unpaid assistants of their fathers, as bankers they brought home the results of their summer's work, in money or goods, making a separate and unique contribution to the family.

Men went to sea because it was what they had been raised to do. Over the years they had acquired the skills to go fishing, and it would be odd if many of them had not gone. Indeed it was almost necessary to be born to the fisheries, and to be involved since boyhood, for very few men brought up in other environments joined the fishing fleets.

While it's easy to conclude there wasn't much that was romantic, or even heroic, about life on the fishing banks, there was nevertheless a certain mystique about the sea, an allure that called men to it.

Evelyn Richardson, who spent thirty-five years living by the ocean at a lighthouse in Shelburne County, put it thus: "Yet while the sea's cruelty repelled men, its beauty drew them....And many a seaman has tried to impress this truth upon his sons. They might as effectually warn the young man of sea-going blood not to look at the pretty girls, lest they prove heavy on the pocketbook and on the heart. To such, the sea's voice has always been irresistible and, like love's, had nothing to do with logic."[7] Thomas Raddall, in his 1951 foreword to W. R. MacAskill's *Lure of the Sea: Leaves from My Pictorial Log*, alluded to the sea's seductiveness: "Ay the lure of it. The bonny face of it that smiles like a sweetheart when you've been far away inland....The sleek and flexuous body of it that's like the swell of breast and hip in the only woman you ever really loved...all a woman's grace in the white thigh of a jib, in the curve of a fore-and-aft mainsail on the windward reach, in the gently arching belly of a topsail..."

Writers could indeed wax poetic about the sea and its seductive qualities. "The sea!" wrote Frederick William Wallace in his novel *Blue Water*:

> The great stretches of clear, clean water rolling tireless and ever changing under the lash of the great untrammelled winds. Always beautiful; ever mysterious; charged with the compelling grandeur of the gale; languorous with the calm and the soft zephyrs of summer, and sublime, yet fearful, in the storm. Swept with the winds of heaven, which, laden with brine, fill the lungs with their cleansing purity and clear the brain befuddled with the vice and dissipations of the land, the mighty watery wastes purge body and soul and rejuvenate the jaded mind....As the great rejuvenator, it eradicates the dregs of vice and bestiality hereditary to the land, and makes men of those who fare their lives upon its mighty breast.

In this passage "the vice and dissipations" belong to the schooner's rough and tough crew who've been restored to sobriety and usefulness after several days of drunken revelry in the fleshpots of Gloucester. But many did believe that the sea had restorative powers, the ability to change men for the better.

Many men were "committed to the toil of the sea," drawn irresistibly to it. The vessels they sailed in went down, or grew old. But the sea abides. "The ribs of many schooners, slimy and rotten," wrote Norman Duncan in *The Way of the Sea*, "and the white bones of men in the offshore depths, know of its strength in that hour—of its black, hard wrath, in gust and wave and breaker. Eternal in might and malignance is the sea! It groweth not old with the men who toil from its coasts. Generation upon the heels of generation, infinitely arising, go forth in hope against it, continuing for a space, and returning spent to the dust....As it is written, the life of a man is a shadow, swiftly passing, and the days of his strength are less; but the sea shall endure in the might of youth to the wreck of the world."

Fishing on the banks was a masculine occupation. After quoting the old saying that a man who would go fishing for a living would go to hell for a pastime, Rear-Admiral H. F. Pullen in *Atlantic Schooners* wrote, "there is no doubt that the great fishing banks of the Northwest Atlantic bred a race of seamen unexcelled in hardiness, resourcefulness, courage and seamanship." For many boys and young men, going to sea was a passage to manhood. Ultimately it became a test of manhood. From flunkey to doryman they were learning to become men, "all blooded Bankers...hardy men an' bold," as Frederick William Wallace put it in *The Shack Locker*. Watching young fishermen scrubbing the hull of a returned vessel and repairing dories on a Lunenburg dock, the American writer T. Morris Longstreth, in *To Nova Scotia: The Sunrise Province of Canada*, enthused: "Most of the hands were boys in the first bronzed glow of manhood, thick of neck and arm, moving with that slow and easy grace which seemed to have caught and individualized the motion of the sea. I was reminded, by some trick of memory, of the cowboys I had watched working outside corrals a continent away." Novelist Hugh MacLennan was also moved by the sight of sturdy fishermen on a Lunenburg wharf, and remarked to his wife Dorothy Duncan that some Nova Scotians thought that only fishermen deserved the name "Bluenose": "Look at the muscles on those fellows straightening out the tackle over there. Some people in Nova Scotia say none of us should be called Bluenose but those fishermen. That's nonsense, of course. But these people are certainly more colorful than any of the rest of us."[8]

During the war, in February 1917, the *Canadian Fisherman* noted that many patriotic fishermen in the past year had enlisted for service overseas. While admirable, the journal maintained, if the enlistments continued the industry would be in serious trouble. Only men of good physique and courage were able to engage in the dangerous calling of fishing; it was no work for weaklings, and women were simply unable to do it. "It is well understood that this breed of manly men are ready and willing to do their bit but their bit may consist of procuring food for those at the front and is every bit as important as that of shouldering a rifle."

At the best of times, life aboard the fishing schooners and in the dories was hard. Rough and dangerous. And dirty. There was a feeling of camaraderie among the men, a sense of brotherhood. As fishermen they shared a common experience, working together, facing the same risks, attempting to make a decent living to provide for their families on shore. Historian M. Brook Taylor in *A Camera on the Banks* described the insouciant attitude of the men towards their way of life: "In this masculine world of work, questions and complaints were suppressed, humour and endurance encouraged. Captains and crews bragged of the risks taken and overcome, the dangers barely survived, the runs to port in the teeth of adverse conditions." Many were the tales of shared endurance, courage, and tragedies at sea.

MacAskill: Seascapes and Sailing Ships, published in Halifax in 1951, is a collection of Wallace MacAskill's most powerful visual images of Nova Scotia. Among the more than one hundred photographs included, Ian McKay remarked in *The Quest of the Folk*, there are rugged, hardy men everywhere, "braving the sea, clamping down picturesquely on their pipes, instructing their sons in nautical lore, building their wooden ships, marching with oxen, unloading fishing vessels." It was a man's world. Not a woman in sight.

What was the role of women in the world of fish?

When schooners arrived home from the banks with a load of salted cod, it was women, largely, who "made" it, that is, cured the fish by spreading it to dry on flakes along the shore or, notably in Grand Bank, Newfoundland, on the beaches. It was hard, backbreaking work, with long hours and low pay. The fish had to be turned regularly, and protected from the elements. In his history of Newfoundland and Labrador, *As Near to Heaven by Sea*, Kevin Major observed that Grand Bank "beach women" worked from daylight to dark, much of the day bent over, even if pregnant, often with only one chance to get home to prepare meals for their families. Garfield Fizzard told the story of one woman who worked all summer on the beach making fish, starting at five o'clock in the morning, for thirteen dollars (she was paid ten cents a quintal). Settling up in the fall, the owner said: "You still owes me two cents." She managed to get the two cents from the captain of the vessel.

While the curing of fish was one important way that women took part in the prosecution of the fishery, according to historian Suzanne Morton they were seen by both their communities and the state primarily as the wives and daughters of fishermen. "Although the paid and unpaid shore work of women in Atlantic fisheries was unquestionably essential to the industry's operations, it was almost always disregarded, minimized, or ignored." In the end, women were considered most important as housewives, as consumers who would increase the demand for fish as food.[9]

Even if not involved directly in the fishery, the life of a fisherman's wife was not an easy one. While the men were away on the banking vessels, the women were the backbone of the family, for they did all the work at home. As well as bearing and rearing many children, they looked after gardens and harvested crops that were sometimes put in by the men before they went to sea. Women hauled firewood and water, they made bread, they made soap and washed and mended the clothes. They also took care of the animals. They milked the cows and slaughtered the pigs. And they picked berries. The fishermen were able to make ends meet because their wives worked so hard at home while they were out to sea. As Jim Pittman, Lunenburg's "singing cook," put it, the fisherman's wife was the best catch of his life!

The women also had to watch and wait for their men to come home, knowing that some would never return. If widowed they often faced a life of great hardship and penury. It was their lot in life to watch for the sails of returning schooners, hoping against hope that the flag would not be flying at half-mast.

The role of women in the fishery is finally being recognized. At the mariners' memorial in Grand Bank, unveiled in August 2007, the names of the 350 or so men lost from the town are inscribed on individual stainless steel plaques submerged under water in a shallow pool covered with beach rocks. Overlooking the pool is a bronze statue of a woman standing steadfast, alone in a widow's walk, her skirt blowing in the wind, meant to symbolize the virtues and strength of character of the thousands of Newfoundland and Labrador women who have lost men to the sea. Two monuments grace the waterfront in Gloucester. The fishermen's memorial cenotaph, known as "The Man at the Wheel," unveiled in 1925, features a fisherman in sou'wester and oilskins clutching the wheel of his schooner while bracing himself against the sloping deck; below the bronze statue, on the stone pedestal, is an inscription from the 107th Psalm: "They that go down to the sea in ships—1623–1923." Further along the esplanade there is now another statue, a woman with two children facing seaward, with an inscription around the base: "The wives, mothers, daughters, and sisters of Gloucester fishermen honor the wives and families of fishermen and mariners everywhere for their faith, diligence, and fortitude." The sculpture, commissioned by the Gloucester Fishermen's Wives Association, was unveiled in August 2001 to honour the women who have been "the soul of fishing communities." Angela Sanfilippo, GFWA president, said at the dedication ceremony: "The memorial serves as a testimonial to what wives, mothers, sisters, and children of fishermen of the world have endured because their men chose to be on the water. They had no choice but to stand on rock, to be on land."

In *Blue Water* Wallace reflected on the fate of the protagonist's mother, and fishermen's wives in general:

> [B]orn and bred in a seafaring community, with the sea ever before her eyes and the breath of it in her nostrils, she knew that the day was coming when her only child would answer to the call. It was hard, but the womenfolk of that sea-washed coast were steeled to bear it; nerved to endure the racking anxieties of days and nights when the great Atlantic combers thundered in acres of foam upon the iron rocks, and the grey scud flew low before the spite of the gale; when the spindrift froze in the biting air and slashed through the howling snow-grey nights when wind and sea arose in rage and shrieked for victims. God! They need His help when their men are at sea!

THE CALL OF THE SEA

The banks fishermen who lost their lives on the ill-fated schooners in the 1920s were the last generation, in some ways, to confront the most extreme dangers of the all-sail fishing fleets. To be sure, men continued to die in the fisheries thereafter. There are, for example, 240 names on the fishermen's memorial in Lunenburg from 1928, the year after the tragedies, to the end of the twentieth century. But the days when vessels had to rely solely on wind and muscle to keep them off the Sable Island bars were passing.

Much of the improved technology was slow coming, and sometimes seemed to be intended to increase the catch of fish rather than to improve safety. There were plans to introduce radio equipment and engines to fishing schooners after the 1926 gale, but it took the great losses of 1927 to jolt the vessel owners and the government into action. In May 1928 the Canadian Department of Marine and Fisheries began daily noon-hour radio broadcasts from Louisbourg in Cape Breton so that vessels equipped with radio receivers could at last get news from shore, especially weather reports. Within a few years of the gales radios as well as engines became standard equipment aboard schooners. At last there would be some advance warning if a bad storm was heading towards the schooners, though ship-to-shore radio did not come to vessels in the Lunenburg fleet until 1933.

When the loss of *Columbia* was confirmed in the fall of 1927, her owner Captain Ben Pine had declared that he would never again send a craft fishing unless she had an engine. Change had begun around the beginning of the twentieth century when in March 1900 Captain Solomon Jacobs commissioned the *Helen Miller Gould*, "her arrival set[ting] in motion a process that would transform the fleet in less than 20 years." The new auxiliary vessels, wrote John Morris, were propelled by the traditional fore-and-aft sails plus a gasoline-powered, propeller-driven engine. There were plenty of problems with the early powered vessels—explosions, breakdowns, and fires, including the one that burned the *Gould* to the waterline at North

Sydney, Nova Scotia, in October 1901—and they were also expensive to run, requiring, for example, an engineer. But eventually they would replace the sailing schooners, and be replaced in their turn by all-power steamers.

Though it was many years before auxiliary engines were widely used, especially on Lunenburg and Newfoundland vessels, nevertheless they were eventually adopted, and schooners no longer had to rely on canvas alone in an emergency. Life aboard was also made easier by the installation of gasoline engines on deck, used to hoist sails and weigh anchors, the first such engine in the Lunenburg fleet appearing on the schooner *Leta J. Schwartz* in 1912. "Gasoline engines revolutionized the fisheries," said W. Jeffrey Bolster in *The Mortal Sea*. "They offered security, efficiency, and ease from the backbreaking labor of rowing or handling sail. And like the answer to a prayer, they provided headway in frustrating calms and frightening gales."

A serious problem in the past had been the loss of many family members in the same shipwreck. Following the disasters two years in a row, according to Ralph Getson, the fishing companies in Lunenburg issued a directive that there could be no more than one family member on board a vessel at the same time. The practice continued, but nothing like in the past.

An important technological development was the invention of the knockabout fishing schooner by Thomas F. McManus of Boston. Having often looked at the footrope under the bowsprit on which men stood to take in the jib, McManus determined to find a way to eliminate the dangerous spar, nicknamed the "widow maker" by fishermen. In rough weather, or when the footrope became thick with ice, the men, bundled in their awkward oilskins, often lost their footing and were swept away. Yet the bowsprit was "the linchpin of a sailing vessel's standing rigging." McManus's brilliant design rounded the bow and did away with the bowsprit, allowing the men to work the headsails from the relative safety of the deck.

The first knockabout, the *Helen B. Thomas*, was built by the Oxner & Story shipyard on the Essex River near Gloucester over the winter of 1901–1902. The second knockabout, the *Alcyone*, which fished out of Digby, was built by Amos Pentz, the master shipwright of McGill shipyard in Shelburne, in 1904. The first knockabout built in Lunenburg was the *General Haig*, launched in 1919. In his later semi-knockabout design, McManus lengthened the round bow of the knockabout in order to shorten, but not eliminate, the bowsprit; the first semi-knockabout, the *Mooween*, was launched at Gloucester in early 1904. The fact that men on sailing vessels no longer needed to go out on the bowsprit to manage the foresails, combined with auxiliary engines, made the vessels considerably safer. As W. M. P. Dunne wrote in his biography of McManus, "In the years between 1910 and 1930, when the swan song of fishing schooners resounded over New England waters, it was not so much a case of sail *versus* power as it was of sail adapting to power." Naval architect and maritime historian Howard I. Chapelle, Dunne said, considered the knockabout "the acme in the long evolution of the New England fishing schooner."

Another technological change was the coming of the steam trawler, or dragger, more important than improvements to sailing ships because the trawler provided a real and lasting alternative to the old way of doing things. Steam trawlers dragging huge nets through the water had been prevalent in European fishing fleets for some time, but now they were slowly gaining ground in North America. Bowring Brothers in Newfoundland had tried steam trawling as early as 1901, but the experiment was not successful. Gloucester's first trawlers were ordered from the shipyard in Essex after the First World War, a decade after they had appeared in Boston. Halifax, the Canso area, and Lockeport were the first ports to use trawlers in Nova Scotia, providing produce for the increasingly important fresh fish market.

Many fishermen all along the shore were vehemently opposed to the big new steam trawlers, believing, all too true as it turned out, that they would harm the fish stocks and take their jobs. In Newfoundland the Union Party, the political wing of the Fishermen's Protective Union, expressed strong reservations about what it regarded as the destructive nature of dragger technology. In Gloucester, according to Mark Kurlansky in *The Last Fish Tale*, the Master Mariners Association argued in 1911 that "the continued operation of these trawlers scraping over the fishing grounds and destroying countless numbers of young and immature fish, is the greatest menace to the future of the fisheries, the greatest danger the fisheries have ever faced along this coast." That same year Canadians and Newfoundlanders called for a ban on the use of steam trawlers that were seen to be threatening young fish and their habitat as well as wrecking the fishermen's gear.

But the trawlers were the wave of the future. Michael Wayne Santos noted in *Caught in Irons* that where once a forest of masts had dominated Gloucester's waterfront, the two-masted schooner was an oddity by the 1920s; many old schooners had been sold and others converted to draggers by having their rigs cut down and engines installed. Certainly the banks schooners lingered on much longer in Lunenburg and Newfoundland, but the onslaught of the trawlers would continue and ultimately prevail. The *Canadian Fisherman* of November 1927 suggested that the steam trawlers, "large, sturdy capable vessels...came as inevitably and naturally as tractors to the farms of Canada." Colin McKay in the December 1927 issue argued that trawlers were a more efficient way of fishing and likely safer, but it was natural that fishermen, fish merchants, and others would resent the intrusion. The old order had provided much employment for "a multitude of skilled craftsmen," from blacksmiths to riggers to sailmakers. It was sad that the upsetting of the traditional environment would shake people "from the tree of life like ripe fruit," but that was "one of the tragedies of progress." Besides, if trawlers were banned from Canadian ports as some desired, they would just be based in Gloucester or Boston and sell their fish into the Canadian market.

In *Witch in the Wind* Marq de Villiers pointed out that the changes were particularly hard on the schooners' skippers, for it made their skills irrelevant. "No point in learning to think like a fish, or to understand the sea, or to have a feel for your living vessel and her skittish ways....All you had to do now is drag a net or a bucket along the bottom, catching everything that swims and ruining everything you touched...." Angus Walters, in a statement both sexist and ageist, agreed: "In those days you had to know something. Today you could put an old woman into 'em."

In *The Mortal Sea*, Bolster commented that in the 1920s the schooners and the steam trawlers shared the banks. "The proximity of staunch schooners (whose designers and builders were still making refinements) and modern trawlers (which fished with clockwork precision) signalled that the age of sail was not yet finished, though it was passing rapidly…painters and photographers still found the fleet romantic, heroic, and wistfully traditional." Like almost every other aspect of life at the beginning of the roaring twenties, he said, the fisheries were modernizing.

A Newfoundland emigrant in New England in 1950, looking back nostalgically at the country he had left behind, evoked a vision of a kind of golden age: "We envision tall-sparred schooners, white sails spread taughtly for spring airing, mirrored in the harbour calm. At least, that's the picture we are likely to recall. Most of us in the States came here before the high bars went up on the U.S. border in the early '30s. We can't picture the modern craft, which are mainly mechanized hulls sans canvas, sans romance. Anyway, who wants to dream about a steam trawler?"[10]

> The armada of canvas is hull-down over time's horizon.
> —Joseph Garland, *Gloucester on the Wind*

Arthur Hunt Chute wrote in his novel *The Crested Seas*: "Hideous, foul-smellin' gas trawlers will soon replace the tall white schooners. When they pass, our last real sailors will have gone to keep company with the clipper captains and the packet rats. When the liners no longer hail her from the ocean lanes, folks will doubtless pause in art galleries and museums to marvel at this loveliness which while here was hardly heeded.... We never miss the beauty, until the beauty's gone."

It was the end of an era. In Gloucester, Gorton's last working schooner, the *Thomas S. Gorton*, built in 1905, sailed until 1956. The Newfoundland banks fishery continued up to the time of the Second World War, the last schooner constructed in Grand Bank in 1935.[11] In Lunenburg the *Theresa E. Connor*, "the last of the salt bankers," was built in 1938. She was supposed to be called the "Nellie E. Connor," the nickname of the wife of the president of the Maritime National Fish Company of Halifax. However, Captain Clarence Knickle, who was to command her, objected to the name having thirteen letters. The launch was delayed a day as the name was painted over and the new one put on. According to the *Halifax Herald*, the launch was also held up because the builders at Smith & Rhuland in Lunenburg had qualms about launching their 181st vessel on the 13th day of December. Superstitions endure. In May 1963 Captain Harry Oxner took the vessel to Fortune Bay, Newfoundland; unable to find the additional crew he needed, the *Theresa E. Connor* became the last of the saltbank schooners to sail out of Lunenburg.

The schooners and the men in the fishing industry who built the towns they sailed from are gone, but a few of the vessels do remain. In 1967 the *Theresa E. Connor* was made the centrepiece of the Fisheries Museum of the Atlantic at Lunenburg; a photograph of her on the Lunenburg waterfront, along with two other vessels, taken in 1939, with the world on the brink of war once again, ended up gracing the Canadian one hundred dollar bill from 1976 to 1990. The *Sherman Zwicker*, built at Smith & Rhuland shipyard in Lunenburg in 1942 for Zwicker & Company, sailed for many years out of Boothbay Harbor, Maine. The *Adventure*, "the last of the Gloucestermen," a knockabout schooner with an auxiliary engine, was built at Essex in 1926; when she retired in 1953 she was the last of the American dory fishing trawlers left in the Atlantic. In early September 2012 the restored *Adventure* led the Joseph E. Garland Parade of Sail in Gloucester, and there are plans for her to sail at sea once more. The spirit of *Columbia* may also sail again, for a replica schooner is being built in Panama City, Florida, with spars, rigging, sails, blocks, and metal work provided by companies in and around Lunenburg. The *Bluenose II*, built in Lunenburg as a replica of the original *Bluenose* in 1963, was rebuilt and relaunched in 2012; in the words of Peter Carnahan in *Schooner Master*, "She sails the Atlantic waters like a constitutional monarch, dignified, unchallenged, expensive, a locus of pride and nostalgia."

An old fisherman in Peter Barss's *A Portrait of Lunenburg County* perhaps best summed up the ending of the great era of the Lunenburg banks fishing fleet: "Well after them August Gales, it just seemed that…from then on vessel fishin' seemed to die out. I wouldn't just like to say that it was on account o' the August Gales that vessel fishin' died out. It could'a been somethin' like that…I don't know…after a while the men wouldn't go. They couldn't get a crew….It kept droppin'…till they was all gone. An' they started to get these here draggers…draggers started to take over. Easier work, you understand. A lot o' vessels was sold up in Newfoundland an' a lot was lost. They was sold or lost. There's one yet up here in Lunenburg—the *Theresa E. Connor*—but she's only a museum. That's all that's left."

AFTERWORD: DAYS *of* FESTIVITY *and* HOPE

THE ELEVENTH ANNUAL LUNENBURG FISHERMAN'S REUNION AND PICNIC was held on September 22, 1927. The memorial service to honour and mourn the loss of the eighty-four men who had died on the four Lunenburg schooners off Sable Island in August would be held later, in a couple weeks' time. But, for now, it was time to think of the living. The fishermen had returned from their long summer voyage, and were able to spend time with their families.

Some four thousand people had attended the tenth annual fishermen's picnic in Lunenburg in September 1925, according to the *Canadian Fisherman* in November that year. In 1926, however, with the loss of the *Sylvia Mosher* and the *Sadie A. Knickle*, it was decided that such frivolity would be out of place. But attitudes were different in 1927. Perhaps people felt they could only mourn so much, that it was time to look to the future.

"Another fishing season is ended for the Lunenburg deep sea fishermen," announced the *Bridgewater Bulletin* on September 27, "and again the town is in gala appearance to welcome home these toilers of the deep." Crowds were flocking into the town from around the county and from distant parts of the province, incoming trains were packed, the hotels were full of people and garages were full of autos. Some of the crowd, estimated at four thousand, also arrived in hundreds of small boats.

The *Bulletin* remarked that only one thing marred the success of the picnic: the fact that more than 130 men had been lost since the last picnic in 1925. The families left behind could take no part in the pleasure of the day. Notwithstanding this year's disaster, however, those in charge had worked indefatigably to make the day a successful and joyous one for those who were fortunate enough to return.

The *Lunenburg Progress-Enterprise* on September 28 pronounced the fishermen's reunion to be "one of the best." The day was ushered in with the ringing of bells and the blowing of horns. At 10:00 A.M., "under blue skies and balmy air," people lined the finger wharves jutting out into the harbour to view the water sports, the first program of the day. There were dory races, a split tub race, and the greasy pole, on which foolhardy young men tried to slide from the wharf to the end of the pole to catch the flag and receive a modest reward. Then at 1:30 the Trades and Calithumpian procession assembled in the grounds of the Academy, and, headed by the Liverpool band, six hundred schoolchildren paraded through the streets, the children representing lilies, rosebushes, little folk from Ireland, applesauce and stringbeans, radios, pirates, gypsies and Indians, and Confederation. Little girls dressed as present-day flappers walked alongside demure young ladies from 1867. The children were followed by gaily decorated cars and floats "of clever originality" by various town companies, headed by the Lunenburg band.

Most spectacular beyond doubt was the Canton Swastika IOOF float, providing a vivid splash of colour with its "beautiful uniforms and dazzling costumes." William Emeneau won first prize for trade floats with a replica of his cooperage, with workmen inside the display busily engaged in pounding out a tattoo on the fish casks. Second prize went to C. D. Ritcey's furniture store for a tastefully

AFTERWORD: DAYS OF FESTIVITY AND HOPE

decorated room with a bevy of ladies enjoying afternoon tea. First prize in the Calithumpian section went to "Mr. and Mrs. Rouse," portraying the first settlers in 1753, "as no parade in Lunenburg is complete without them." Powers Motor Company exhibited a Pontiac car, preceded and followed by "a band of Pontiac Indians gaily decked out in war paint and accoutrements, apparently peaceably inclined in spite of the fact their chief demonstrated his capability of emitting the most blood curdling of war whoops." The Lunenburg Outfitting Company displayed, most appropriately, a huge codfish, but unfortunately, and perhaps symbolically, the vehicle propelling this float broke down.

Store windows were also decorated, and W. N. Zwicker & Company won first prize: on one side seven model ships displayed the work of the veteran shipbuilder, eighty-one-year-old Solomon Morash, and the window on the other side of the door held the cups won by "our champion Bluenose" during her racing career. "Many encomiums heard on all sides." A. Dauphinee & Sons on Montague Street displayed a full-rigged ship surrounded by products of the firm such as ships' blocks.

Field sports were held later in the day on the athletic grounds on Blockhouse Hill. An evening program included a band concert, dancing, distribution of prizes; excitement was high until midnight, when the Dempsey-Tunney fight from Chicago was broadcast on the radio. Prizes were given out for such events as the dory races and the running races, and a notable prize went to Dr. Russell Zinck and Dr. Herbert Himmelman for winning the fat man's race.

A few months later, on January 10, 1928, the *Bridgewater Bulletin* reported on another memorable day in Lunenburg. The day before, the newspaper said, some 200,000 quintals of fish had been sold in the town, with some two million dollars in cash going into the coffers of the Lunenburg fishermen. The waterfront was a scene of great activity, even of congestion, as late into the night all roads leading to the town were crowded with traffic bringing loads of fish to market, some by motor transport but much of it being trundled along the road by the historic ox teams. Thousands more quintals of fish were landed on the piers from small boats that came from along the shore. The fish being marketed was the remainder of the year's catch that had been held back for higher prices. The *Bulletin* declared that the flow of the fish to the central market in Lunenburg was, in its way, to the Nova Scotia fishing industry what the "head of the Lakes" was to the movement of grain in western Canada. Several Lunenburg fish merchants claimed that every one of the ninety vessels in the 1927 fleet had cleared the year's operations with a profit. Lunenburg, they contended, had "the most economically operated fleet in the world."

It was, then, a time of hope, and apparent prosperity. It would be thirty-six years before the *Theresa E. Connor* would limp home from Newfoundland unable to find enough crew members for one final trip to the banks. Uncertain times ahead, perhaps, but still a future for the Lunenburg banks fishing fleet. Plenty more fish to be caught…

ACKNOWLEDGEMENTS

The town of Lunenburg is fortunate to have on its waterfront a very fine institution, the Fisheries Museum of the Atlantic. And the museum is blessed to have Ralph Getson as its curator of education. Apart from his extensive knowledge of the museum and its contents, Ralph is able to direct the expectant researcher unerringly to the most obscure and worthwhile sources in his maritime archival empire. The museum's resources, including the many books in the library, have made this book possible. Ashlee Feener, Damara Mossman, and others have made my many visits to the museum not only useful but fun.

The staff at the Lunenburg branch of the South Shore Public Libraries were particularly helpful, acquiring for me through the Interlibrary Loan Service of Library and Archives Canada the microfilm of the Halifax and St. John's newspapers, essential to this study; sadly, this valuable service has now ended, after sixty years, because of cutbacks by the Harper government. Barbara Spindler and her colleagues at the South Shore Genealogical Society put up with me for many hours as I made my way on microfilm through the pages of the *Canadian Fisherman*. Also in Nova Scotia, Doug Berrigan helped me out at the LaHave Islands Marine Museum, as did Kim Robertson Walker at the Shelburne County Archives and Genealogical Society. Numerous books and sources were viewed at the Killam Memorial Library and in the archives at Dalhousie University in Halifax.

In Newfoundland I had the pleasure of visiting the Burin Heritage Museums, where curator Tina Comby was especially helpful and Wayne Hollett shared his knowledge of the Burin fishery. In Grand Bank Robert Parsons provided tea as well as much useful information. I'm indebted as well to Joan Ritcey and her staff in St. John's at the Centre for Newfoundland Studies at Memorial University.

Among the many people I must thank are Marq de Villiers and Sheila Hirtle, who encouraged me from the beginning and even shared research notes. Others who read and commented on the manuscript, besides Marq, were historians Tina Loo, Suzanne Morton, and M. Brook Taylor (who had convinced me to tackle the subject), my former colleagues at the University of Toronto Press, Ian Montagnes and Jean Wilson, my brother-in-law Andy Gray, and, always my first reader, Michael Browne.

Ralph Getson read each chapter as it emerged, thus saving this former Ontario farmboy (who couldn't tell a jib from a jumbo) from the more outrageous of nautical mistakes. At Nimbus it has been a pleasure to work with my editor Patrick Murphy. Paula Sarson improved the manuscript with her fine editing. Any shortcomings or errors that remain in the text are of course my responsibility.

Robert Parsons and Peter Barss kindly gave me permission to quote from their works. Many old and valuable books were loaned to me by Sheevaun Nelson of Blue Rocks. Her website, Fishing? – It was 'A Way of Life' and Lost at Sea, proved to be an invaluable starting point for the project.

A special thanks to Lunenburg artist Jay Langford, who created the splendidly stormy seascape that graces the cover of this book. Thanks also to my friend Peter Tanner of Blue Rocks, and his gaff-rigged cutter *Borden T.*, for giving me at least some experience sailing on the waters of the North Atlantic.

Anyone who has beheld the dignified solemnity of the many fishermen's memorials in Atlantic Canada and in Gloucester will know that the shipwrecks in the August gales of the 1920s, though certainly tragic, are only among the most dramatic of events in a much larger and terrible story. Fishermen have toiled in the North Atlantic to earn a living for centuries, and thousands have perished over the years. Indeed, as recently as this February five young fishermen from Woods Harbour, Nova Scotia, died when their fishing boat *Miss Ally* foundered in a North Atlantic storm. I hope this book serves to honour the memory of the men who met their fate in the August gales and all the other fishermen who have lost their lives in their efforts to bring us fish from the sea.

G. H.
Lunenburg, NS
August 2013

NOTES

The sources used in the writing of this book are listed in the bibliography that follows these notes. A manuscript with a full set of notes, with page numbers referenced, is on file at the Fisheries Museum of the Atlantic in Lunenburg, and can be consulted there.

1: Lunenburg and the Banks Fishery

This chapter is indebted especially to B. A. Balcom's *History of the Lunenburg Fishing Industry* and to Ruth Fulton Grant's *The Canadian Atlantic Fishery*.

1. Chambers et al., *Historic LaHave River Valley*; Haliburton, *The Old Judge, Or Life in a Colony* (1849), quoted in Innis, *The Cod Fisheries*.
2. Fischer, *Champlain's Dream*.
3. The tonnage is given in Spicer, *The Age of Sail*: his imperial measurements have been converted to metric.
4. On page 56 and following, Taylor gives a good description of the work of dorymen, in this case on a schooner out of Digby.
5. The Tanner letters and Ralph Getson's quotation are from "Fishing?—It was 'A Way of Life' and Lost at Sea," web.archive.org/web/*/http://lostatsea.ca, S. Nelson, 2006. Letters reproduced there courtesy Ruth Wagner McConnell.
6. H. R. Aerenburg, "The Famous Fishing Fleet of Lunenburg," *Canadian Fisherman*, May 1925.
7. Simon Watts, "German Subs Spared the Crews, But Not the Schooners," in *National Fisherman*, January 1979, clipping in the LaHave Islands Marine Museum.

2: An Island in a Stormy Sea

The epigraph at the beginning of this chapter is from "Tomorrow's Tide," in Andy Wainwright and Lesley Choyce, eds., *The Mulgrave Road: Selected Poems of Charles Bruce* (Porters Lake, NS: Pottersfield, 1985).

1. Ed Rappaport, deputy director of the National Hurricane Center in the United States, *St. John's Telegram*, October 2, 2010.
2. Cape Breton newspapers quoted in H. Toynbee, "On the Hurricane of August 1873," *Quarterly Journal of the Royal Meteorological Society*, vol. 2, no. 9 (1875).
3. Thurlow, *The Weather Notebook*, CBC Radio, Charlottetown, August 6, 1998, transcript # 245-4.
4. For Walker's fateful expedition see his entry in the *Dictionary of Canadian*

Biography, vol. 2. Hugh A. Halliday in *The Oxford Companion to Canadian History* claimed the fleet lost its bearings in heavy fog and 950 men drowned; Alan Ruffman, "Hurricane Hortense: A Fortuitous Warning," *Shunpiking*, vol. 2, no. 15 (September 1997), said the 1711 storm was probably a hurricane and that 2,000 lives were lost.

5. Ruffman, "The Multidisciplinary Rediscovery and Tracking of the 'Great Newfoundland and Saint-Pierre et Miquelon Hurricane of September 1775,'" *Northern Mariner/Le Marin du nord*, vol. 6, no. 3 (July 1996). On "the hollies," see Ruffman quoted in Deanna Stokes Sullivan, "The Forgotten Storm," *St. John's Telegram*, October 2, 2010.

6. Adele Townshend, "The Loss of the *Welcome*," *The Island Magazine*, no. 28 (Fall/Winter 1990).

7. Parsons, "Eyewitness in a Killer Storm," in Mysteries of Canada series, www.mysteriesofcanada.com, Bruce Ricketts.

8. The description of the island here and above comes mainly from Zoe Lucas, "Postscript" in Christie, *The Horses of Sable Island*; from de Villiers and Hirtle, *A Dune Adrift*; and from the exhibit on Sable Island in the Fisheries Museum of the Atlantic, Lunenburg. The phrase "place well-known for shipwrecks" appeared in a seventeenth-century book about North America by Johannes de Laet, published in Latin, Dutch, and French, quoted in Campbell, *Sable Island, Fatal and Fertile Crescent*.

9. John D. Reid, "The Great 1900 Galveston Hurricane in Canada," www.magma.ca/~jdreid/great_1900_hurricane.htm. Accessed 2012. Reid wrote that there were eighty-six fatalities in Canada and Newfoundland but considers this a very conservative estimate.

3: The August Gale of 1926

1. Backman, *Bluenose*. For accounts of the April 1926 storm, see Gillespie, *Bluenose Skipper*, and de Villiers, *Witch in the Wind*.

2. In *A Dune Adrift* de Villiers and Hirtle wrote of the 1926 gale that it was "a Force Four hurricane, with winds that circled the compass as it passed through. A strong tide was running towards the island....On the mainland, just a hundred unreachable miles away, it was a beautiful day." More than one source claims it was a fine day on the mainland, but clearly it wasn't pleasant everywhere. Environment Canada said the storm made landfall on the morning of August 8 near Canso as a Category 1 hurricane with winds of 130 kilometres per hour. Since winds are usually stronger on the right side of a hurricane, it's likely the winds were fiercer out to sea.

3. Unidentified newspaper clipping, no date, and notes to Ralph Getson, Fishermen's Museum of the Atlantic, Lunenburg, from Clayton Walter Wamback, September 22, 1987.

4. Rhodes was quoted in the *Halifax Herald*, August 27, 1926. In "Solving

a Problem," Fred Winsor wrote that in the same storm a Norwegian freighter, the *Ringhorn*, bound from Parrsboro for Manchester with lumber, went down near Scaterie Island off the coast of Cape Breton. Five of the ship's crew died trying to reach land. The loss of this foreign freighter gave rise to an inquiry within eleven days, but no inquiry was ever held into the far larger disasters in the Nova Scotia schooner fleet.

5. Literary critic Gwendolyn Davies, in her afterword to *Rockbound*, wrote that Day's imaginative depiction of the sinking of the ship is closely linked to accounts of actual events in the *Halifax Herald* and other newspapers of the time.

6. Email from Charles D. Roach, Centre de généalogie Père Charles Aucoin, Chéticamp, September 1, 2011; *Halifax Morning Chronicle*, August 27, 1926, re Muise's connection to the baron.

7. "Under Sail before Gale," unidentified newspaper clipping, LaHave Islands Marine Museum.

8. "Biography of Percy Baker by Norman McLetchie 1992," LaHave Islands Marine Museum. Herbert Getson was Ralph Getson's great uncle.

9. *Bridgewater Bulletin*, September 12, 1926, April 19, 1927. An account of the event was also give to Ralph Getson by Walter Wamback's daughter, Vivian Corkum.

4: The August Gale of 1927

1. American Meteorological Society, *Monthly Weather Review 1927*, Storms and Weather Warnings, journals.ametsoc.org/loi/mwre. A Category 2 hurricane has winds from 154 to 177 kilometres an hour.

2. Information provided by Captain Corkum's son to Ralph Getson, Fisheries Museum of the Atlantic, Lunenburg, 2007.

3. Robinson, *Men against the Sea*, and Fisheries Museum of the Atlantic exhibit. According to the *Lunenburg Progress-Enterprise* of October 31, 1928, the *Andrava*, Captain Leo Lohnes, presumably the same vessel, went down in the harbour at Sydney that month after being rammed by the French trawler *Commandant Emaille*. The schooner sank in about five minutes, but the crew of twenty-one was saved.

4. Length and tonnage of the vessels here and in chapter three come from *List of Shipping issued by the Department of Marine and Fisheries being a list of vessels on the Registry Books of the Dominion of Canada* (Ottawa: King's Printer, 1925 and 1926); the measurements have been converted into metric. Of the six Lunenburg vessels that foundered, only the *Sylvia Mosher* was not found in this source.

5. Archibald Enslow provides a fine example of how confusing the research on the lost men can be. His name does not appear on the Lunenburg fishermen's memorial, or among the names at the Fisheries Museum of the Atlantic, or in the crew list printed in the *Shelburne Gazette and Coast Guard*, September 8 and 15, 1927. However, it does appear in a

list in the *Coast Guard* on November 3, in *Lost Mariners of Shelburne County*, vol. 1, and on the fishermen's memorial in Shelburne.

    As the *Halifax Morning Chronicle* reported on September 5, records were not kept when the vessels sailed, so it was difficult to be precise about just who was on board. Ralph Getson also explained that there were no proper crew lists for the vessels that foundered: names were taken from insurance papers and store ledgers.

6. Erin Anderssen reports, *Globe and Mail*, Toronto, August 18, 2007.
7. Tragedy struck Blue Rocks again in 1942 when the Lunenburg auxiliary fishing schooner *Flora Alberta* was struck down in dense fog at 2:00 A.M. on the Western Bank fishing grounds, cut in two by the British steamer SS *Fanad Head*, sailing in convoy. Of the 28 men aboard, 21 died, 6 from Blue Rocks. See Spicer, *Age of Sail*, and Supreme Court of Canada, *Fanad Head* v. *Adams* [1949] S.C.R. 407, scc.lexum.org.
8. The Tanner letters are from "Fishing? — It was 'A Way of Life' and Lost at Sea," web.archive.org/web/*/http://lostatsea.ca, S. Nelson, 2006.

5: Newfoundland and the August Gale of 1927

    The principal sources for this chapter are Raoul Andersen's *Voyage to the Grand Banks*, Otto Kelland's *Dories and Dorymen*, and the several books by Robert C. Parsons listed in the bibliography.

1. The line comes from Charles Kingsley's poem "The Three Fishers."
2. On the *John C. Loughlin* see Robert Parsons, "Red Harbour: Death on the Crosstrees," *Newfoundland Quarterly*, vol. 88, no. 4 (Summer 1994); Parsons, *Toll of the Sea*; Brennan letter: "Ship Cove, Cape Shore, via Placentia to Mr. Locklan," Burin Heritage Museums, Burin. On the *Annie Healy*, see Darrell Duke, *Thursday's Storm: The August Gale of 1927* (Flanker Press, 2013).
3. These names are taken from the Port Elizabeth Island Upper Cemetery memorial headstone to the men lost, found on Newfoundland's Grand Banks Genealogical and Historical Data online, ngb.chebucto.org/. Flat Island was renamed Port Elizabeth in 1952.
4. Halpert and Story, eds., *Christmas Mumming in Newfoundland*; Alexander, "Newfoundland's Traditional Economy and Development to 1934," in Hiller and Neary, eds., *Newfoundland in the Nineteenth and Twentieth Centuries*; MacNutt, *The Atlantic Provinces*.
5. Mason quoted in O'Flaherty, *The Rock Observed*.
6. Harris, "Introduction," in Candow and Corbin, *How Deep Is the Ocean?*
7. Rogers quoted in Forbes and Muise, *The Atlantic Provinces in Confederation*; Perret quoted in Innis, *The Cod Fisheries*.
8. James E. Candow, "'Recurring Visitations of Pauperism': Change and Community in the Newfoundland Fishery," in Candow and Corbin, *How Deep Is the Ocean?*; Ryan, *Fish Out of Water*, *Journals of the House of*

Assembly, 1915, Appendix.

9. Written in 1947 by Otto P. Kelland (1904–2004) of Flatrock, Newfoundland, wtv-zone.com/phyrst/audio/nfld/24/mary.htm (© Jocelyn Kelland).

10. Cadigan, *Newfoundland and Labrador; Canadian Fisherman*, July 1914.

11. The information about the losses on the *Joyce M. Smith* comes mainly from the Burin Heritage Museums. Two men from Pool's Cove, Fortune Bay, James and Murdock Hancock, may also have been on the vessel when she went down. Different sources give different information.

12. Kelland, *Dories and Dorymen*. The story of the Walsh tragedy is also told by Barbara Walsh in *August Gale: A Father and Daughter's Journey into the Storm* (Guilford, Conn.: Globe Pequot Press, 2012). Maura Hanrahan's *The Doryman* (St. John's: Flanker Press, 2003) is a fine fictionalized account of the 1935 storm, based on first-hand accounts and historical records.

6: The Gloucester Connection

The principal sources for this chapter are Joseph Garland's *Down to the Sea*, John N. Morris's *Alone at Sea*, J. B. Connolly's *The Port of Gloucester*, Mark Kurlansky's *The Last Fish Tale*, Marq de Villiers's *Witch in the Wind*, Keith McLaren's *A Race for Real Sailors*, and Dana Story's *Hail Columbia!* The epigraph at the beginning of the chapter comes from McAveeny, *Kipling in Gloucester*.

1. Procter quoted in Garland, *Down to the Sea*; Hayden, "Introduction," in *Down to the Sea*.

2. McAveeney, *Kipling in Gloucester*.

3. Fischer, *Champlain's Dream*.

4. "Out of Gloucester," www.downtosea.com, R. Sheedy.

5. Robertson, *Shelburne and the Gloucestermen*.

6. Journals of the Assembly, Nova Scotia, 1837, in Innis, *The Cod Fisheries*; Bebb, *Quest for the Phantom Fleet*.

7. Bebb, *Quest for the Phantom Fleet*.

8. National Archives, Washington, DC: Archives II Textual Reference (Civilian), College Park, MD; Series from Record Group 22: Records of the U.S. Fish and Wildlife Service, 1868–2008, Department of Commerce, Bureau of Fisheries, p 45, container # 15, Privileges of American vessels at Canadian Ports, 3 April 1918, Commissioner of Fisheries to Secretary of Commerce (information from Suzanne Morton).

9. McLaren, *A Race for Real Sailors*.

10. *Times* quoted in Story, *Hail Columbia!* Most of the stories that follow about the men from Shelburne come from Smith, ed., *Lost Mariners of*

Shelburne County, vol. 1.

11. Connolly in Church, *American Fishermen*.

7: The Call of the Sea
The epigraph at the beginning of this chapter is from E. J. Pratt's long poem *Rachel: A Sea Story of Newfoundland in Verse* (1917), in *E .J. Pratt, Complete Poems*, Part 1, ed. Sandra Djwa and R. G. Moyles (Toronto: University of Toronto Press, 1989).

1. Orwell in Denys Blakeway, *The Last Dance: 1936, The Year Our Lives Changed* (2010; London: John Murray, 2011).
2. Unidentified clipping, LaHave Islands Marine Museum; Ralph Getson re Nita Himmelman; Frank in Forbes and Muise, *The Atlantic Provinces in Confederation*.
3. Lines of poetry from Dana, *Two Years before the Mast*.
4. Reinhardt, who was born on Bell Island, in conversation with Simon Watts, LaHave Islands Marine Museum.
5. Parker, *Historic Lunenburg*; de Villiers, *Witch in the Wind*; Cameron in Stanton, *We Belong to the Sea*.
6. Wallace, "Life on the Grand Banks: An Account of the Sailor-Fishermen Who Harvest the Shoal Waters of North America's Eastern Coasts," *National Geographic Magazine*, vol. 40, no. 1 (July 1921); below, Edin C. Hill, in the *New York Sun*, reprinted in the *Lunenburg Progress-Enterprise*, August 29, 1928.
7. Richardson, *Where My Roots Go Deep: The Collected Writings of Evelyn Richardson*, in Stanton, *We Belong to the Sea*.
8. MacLennan quoted in McKay, *The Quest of the Folk*.
9. Morton, "'The End Man Is a Woman': Women, Fisheries, and the Canadian State in the 20th Century," in Guildford and Morton, eds., *Making Up the State*.
10. Overton, *Making a World of Difference*, quoting Ron Pollett, "Summer Madness," *Atlantic Guardian* (1950).
11. *Decks Awash*, September–October 1989.

BIBLIOGRAPHY

Allaby, Eric. *"The August Gale": A List of Atlantic Shipping Losses in the Gale of August 24, 1873*. Saint John: New Brunswick Museum, August 1973.

Andersen, Raoul. *Voyage to the Grand Banks: The Saga of Captain Arch Thornhill*. St. John's: Creative, 1998.

Armstrong, Bruce. *Sable Island*. Toronto: Doubleday, 1981.

Backman, Brian and Phil. *Bluenose*. Toronto: McClelland & Stewart, 1965.

Baehre, Rainer K. *Outrageous Seas: Shipwrecks and Survival in the Waters of Newfoundland, 1583–1893*. Ottawa: Carleton University Press, 1999.

Balcom, B. A. *History of the Lunenburg Fishing Industry*. Lunenburg Marine Museum Society, 1977.

Barss, Peter. *A Portrait of Lunenburg County: Images and Stories from a Vanished Way of Life*. 1978; Halifax: Nimbus, 2012.

Bebb, James T. *Quest for the Phantom Fleet*. Lockeport: self-published, 1992.

Bell, Winthrop Pickard. *The "Foreign Protestants" and the Settlement of Nova Scotia: The History of a Piece of Arrested British Colonial Policy in the Eighteenth Century*. Toronto: University of Toronto Press, 1961.

Bellerose, George. *Facing the Sea: The People of Tancook Island*. Halifax: Nimbus, 1995.

Bolster, W. Jeffrey. *The Mortal Sea: Fishing the Atlantic in the Age of Sail*. Cambridge, Mass., and London: Belknap Press of Harvard University Press, 2012.

Bradley, Wendell P. *They Live by the Wind: The Lore and Romance of the Last Sailing Workboats: The Grand Bank Schooners, the Square-Rigged Training Ships, the Chesapeake Oysterboats, the Fishing Slops of the Bahamas*. New York: Alfred A. Knopf, 1969.

Bruce, Harry. *An Illustrated History of Nova Scotia*. Halifax: Province of Nova Scotia and Nimbus, 1997.

Cadigan, Sean T. *Newfoundland and Labrador: A History*. Toronto: University of Toronto Press, 2009.

Cameron, Silver Donald. *Schooner: Bluenose and Bluenose II*. Toronto: Seal Books, 1984.

Campbell, Lyall. *Sable Island, Fatal and Fertile Crescent*. Hantsport, NS: Lancelot, 1974.

———. *Sable Island Shipwrecks: Disaster and Survival at the North Atlantic Graveyard*. Halifax: Nimbus, 1994.

Candow, James E., and Carol Corbin, eds. *How Deep Is the Ocean? Historical Essays in Canada's Atlantic Fishery.* Sydney: University College of Cape Breton Press, 1997.

Carnahan, Peter. *Schooner Master: A Portrait of David Stevens.* Halifax: Nimbus, 1989.

Chambers, Sheila, Joan Dawson, and Edith Wolter. *Historic LaHave River Valley: Images of Our Past.* Halifax: Nimbus, 2004.

Christie, Barbara J. *The Horses of Sable Island.* Lawrencetown Beach, NS: Pottersfield, 1995.

Church, Albert Cook, with text by James B. Connolly. *American Fishermen.* New York: Bonanza, 1940.

Chute, Arthur Hunt. *The Crested Seas.* 1928; Halifax: Formac, 2012.

Connolly, James B. *The Book of the Gloucester Fishermen.* New York: John Day, 1927.

———. *The Port of Gloucester.* New York: Doubleday, Doran and Company, 1940.

Dana Jr., Richard Henry. *Two Years before the Mast: A Personal Narrative of Life at Sea.* 1840; New York: Bantam Books, 1963.

Day, Frank Parker. *Rockbound.* 1928; Toronto: University of Toronto Press, 1989.

DesBrisay, Mather Byles. *History of the County of Lunenburg.* 1895; Belleville, Ont.: Mika, 1980.

de Villiers, Marq, and Sheila Hirtle. *A Dune Adrift: The Strange Origins and Curious History of Sable Island.* Toronto: McClelland & Stewart, 2004.

de Villiers, Marq, *Witch in the Wind: The True Story of the Legendary Bluenose.* Toronto: Thomas Allen, 2007.

Dickson, Frances Jewel. *Skipper: The Sea Yarns of Captain Matthew Mitchell.* Lawrencetown Beach, NS: Pottersfield, 2009.

Duncan, Norman, "The Fruits of Toil," in *The Way of the Sea.* Ottawa: Tecumseh Press, 1982.

Dunne, W. M. P. *Thomas F. McManus and the American Fishing Schooners: An Irish-American Success Story.* Mystic, Conn.: Mystic Seaport Museum, 1994.

Emanuel, Kerry. *Divine Wind: The History and Science of Hurricanes.* New York: Oxford University Press, 2005.

Fergusson, Charles Bruce, ed. *The Diary of Adolphus Gaetz.* Halifax: Public Archives of Nova Scotia, 1965.

Fischer, David Hackett. *Champlain's Dream.* 2008; Toronto: Vintage Canada, 2009.

Fizzard, Garfield. *Master of His Craft: Captain Frank Thornhill.* Grand Bank Heritage Society, 1988.

———. *Unto the Sea: A History of Grand Bank*. Grand Bank Heritage Society, 1987.

Forbes, E. R., and D. A. Muise, eds. *The Atlantic Provinces in Confederation*. Toronto: University of Toronto Press, 1993.

Fraser, Chelsea. *Heroes of the Sea*. New York: Thomas Y. Crowell, 1924.

Garland, Joseph E. *Down to the Sea: The Fishing Schooners of Gloucester*. Boston: David R. Godine, 1983.

———. *Gloucester on the Wind: America's Greatest Fishing Port in the Days of Sail*. Charleston: Arcadia, 1995.

Getson, Heather-Anne. *Bluenose: The Ocean Knows Her Name*. Halifax: Nimbus, 2006.

Gillespie, G. J. *Bluenose Skipper: The Story of the Bluenose and Her Skipper*. 1955; Fredericton: Brunswick Press, 1964.

Grant, Ruth Fulton. *The Canadian Atlantic Fishery*. Toronto: Ryerson Press, 1934.

Guildford, Janet, and Suzanne Morton, eds., *Making Up the State: Women in 20th-Century Atlantic Canada*. Fredericton: Acadiensis Press, 2010.

Gwyn, Richard. *Nation Maker: Sir John A. Macdonald: His Life and Times, vol. 2: 1867-1891*. Toronto: Random House Canada, 2011.

Halpert, Herbert, and G. M. Story, eds. *Christmas Mumming in Newfoundland: Essays in Anthropology, Folklore, and History*. Toronto: University of Toronto Press, 1969.

Hewitt, Harry W. "History of the Town of Lunenburg." typescript, n.d.

Hiller, James, and Peter Neary, eds., *Newfoundland in the Nineteenth and Twentieth Centuries: Essays in Interpretation*. Toronto: University of Toronto Press, 1980.

Innis, Harold A. *The Cod Fisheries: The History of an International Economy*. 1940; Toronto: University of Toronto Press, 1954.

Journals and Letters of Colonel Charles Lawrence: Being a day by day account of the founding of Lunenburg, by the Officer in command of the project, transcribed from the Brown manuscript in the British Museum. Halifax: Public Archives of Nova Scotia, 1953.

Junger, Sebastian. *The Perfect Storm: A True Story of Men against the Sea*. New York: HarperPaperbacks, 1997.

Kelland, Otto. *Dories and Dorymen*. St. John's: Robinson-Blackmore, 1984

———. *Beautiful Ladies of the Atlantic*. St. John's: self-published, 1986.

Kurlansky, Mark. *Cod: A Biography of the Fish That Changed the World*. Toronto: Alfred A. Knopf Canada, 1997.

———. *The Last Fish Tale: The Fate of the Atlantic and Survival in Gloucester, America's Oldest Fishing Port and Most Original Town*. New York: Ballantine, 2008.

Longstreth, T. Morris. *To Nova Scotia: The Sunrise Province of Canada*. New York, London: D. Appleton-Century Company, 1935.

Major, Kevin. *As Near to Heaven by Sea*. Toronto: Penguin Canada, 2001.

Marriner, Ernest Cummings. *Jim Connolly and the Fishermen of Gloucester: An Appreciation of James Brendan Connolly at Eighty*. Waterville, Maine: Colby College Press, 1949.

MacAskill, W. R. *Lure of the Sea: Leaves from My Pictorial Log*. Halifax: Eastern Photo Engravers, 1951.

McAveeney, David C. *Kipling in Gloucester: The Writing of* Captains Courageous. Gloucester: Curious Traveller Press, 1966.

MacDonald, Edward. "The Yankee Gale." *The Island Magazine*, no. 38 (Fall/Winter 1995).

———. "The August Gale and the Arc of Memory on Prince Edward Island." *The Island Magazine*, no. 56 (Fall/Winter 2004).

McFarland, Raymond. *The Masts of Gloucester: Recollections of a Fisherman*. New York: Norton, 1937.

McKay, Ian. *The Quest of the Folk: Antimodernism and Cultural Selection in Twentieth-Century Nova Scotia*. Montreal and Kingston: McGill-Queen's University Press, 1994.

McLaren, Keith. *A Race for Real Sailors: The Bluenose and the International Fishermen's Cup, 1920–1938*. Vancouver/Toronto: Douglas and McIntyre, 2006.

McLeod, Robert R., *Markland or Nova Scotia: Its History, Natural Resources and Native Beauties*. Markland, 1903.

MacNutt, W. S. *The Atlantic Provinces: The Emergence of Colonial Society, 1712–1857*. Toronto: McClelland & Stewart, 1965.

Morris, John N. *Alone at Sea: Gloucester in the Age of the Dorymen, 1623–1939*. Beverly, Mass.: Commonwealth, 2010.

Noel, S.J.R. *Politics in Newfoundland*. Toronto: University of Toronto Press, 1971.

Oickle, Vernon. *Disasters of Atlantic Canada*. Edmonton: Folklore, 2007.

O'Flaherty, Patrick. *The Rock Observed: Studies in the Literature of Newfoundland*. Toronto: University of Toronto Press, 1979.

Ommer, Rosemary, and Gerald Panting, eds. *Working Men Who Got Wet*, Proceedings of the Fourth Conference of the Atlantic Canada Shipping Project, July 14–16, 1980, Maritime History Group, Memorial University of Newfoundland.

Overton, James. *Making a World of Difference: Essays on Tourism, Culture and Development in Newfoundland*. St. John's: Institute of Social and Economic Research, 1996.

Parker, John P. *Cape Breton Ships and Men*. McGraw-Hill Ryerson, 1967.

BIBLIOGRAPHY

Parker, Mike. *Historic Lunenburg: The Days of Sail, 1880-1930.* Halifax: Nimbus, 1999.

Parsons, Robert C. *Lost at Sea,* vol. 1. St. John's: Creative, 1991.

———. *Toll of the Sea: Stories from the Forgotten Coast.* St. John's: Creative, 1995.

———. *In Peril on the Sea: Shipwrecks of Nova Scotia.* Lawrencetown Beach, NS: Pottersfield, 2000.

———. *Wind and Water: Sea Tales from around Our Coast.* St. John's: Creative, 2003.

———. *Survivors and Lost Heroes: The Sea from Cape Race to Cape Ray.* Grand Bank: Books of the Sea, 2007.

Paul, Daniel N. *We Were Not the Savages: A Mi'kmaq Perspective on the Collision between European and Native American Civilizations.* Halifax: Fernwood, 2000.

Pierce, Wesley George. *Going Fishing: The Story of the Deep-Sea Fishermen of New England.* 1934; Camden, Maine: International Marine, 1989.

Pope, Peter E. *The Many Landfalls of John Cabot.* Toronto: University of Toronto Press, 1997.

Pritchard, Gregory P. *Collision at Sea.* Hantsport, NS: Lancelot, 1993.

Procter, George H. *The Fishermen's Memorial and Record Book: A List of Vessels and Their Crews, Lost from the Port of Gloucester from the Year 1830 to October 1, 1873.* Gloucester: Procter Brothers, 1873.

Pullen, Rear-Admiral H. F., drawings by Commander L. B. Jenson. *Atlantic Schooners.* Fredericton: Brunswick Press, 1967.

Raddall, Thomas H. *In My Time: A Memoir.* Toronto: McClelland & Stewart, 1976.

Reeves, William George. "'Our Yankee Cousins': Modernization and the Newfoundland-American Relationship, 1898–1910," PhD thesis, University of Maine at Orono, 1987.

Robertson, Lewis V. *Shelburne and the Gloucestermen.* 1977; Shelburne: Loyalist Foods, 1991.

Robinson, Cyril. *Men against the Sea: High Drama in the Atlantic.* Windsor, NS: Lancelot, 1971.

Rose, George A. *Cod: The Ecological History of the North Atlantic Fisheries.* St. John's: Breakwater, 2007.

Ryan, Shannon. *Fish Out of Water: The Newfoundland Saltfish Trade, 1814–1914.* St. John's: Breakwater, 1986.

Santos, Michael Wayne. *Caught in Irons: North Atlantic Fishermen in the Last Days of Sail.* Selinsgrove, Penn.: Susquehanna University Press, 2002.

Smith, Eleanor Robertson, ed. *Lost Mariners of Shelburne County.* Shelburne County Genealogical Society, 1991.

Snow, Edward Rowe. *Great Gales and Dire Disasters*. New York: Dodd, Mead and Company, 1952.

Spicer, Stanley T. *The Age of Sail: Master Shipbuilders of the Maritimes*. Halifax: Formac, 2001.

Stanton, Meddy, ed. *We Belong to the Sea: A Nova Scotia Anthology*. Halifax: Nimbus, 2002.

Story, Dana. *Hail Columbia! The Rise and Fall of a Schooner*. 1970; Gloucester: Ten Pound Island, 1985.

Taylor, M. Brook. *A Camera on the Banks: Frederick William Wallace and the Fishermen of Nova Scotia*. Fredericton: Goose Lane, 2006.

Thomas, Gordon W. *Fast and Able: Life Stories of Great Gloucester Fishing Vessels*. 1952; Beverly, Mass.: Commonwealth Editions, 2002.

Vickers, Daniel, with Vince Walsh. *Young Men and the Sea: Yankee Seafarers in the Age of Sail*. New Haven and London: Yale University Press, 2005.

Villiers, Alan. *Wild Ocean: The Story of the North Atlantic and the Men Who Sailed It*. New York: McGraw-Hill Book Company, 1957.

Wallace, Frederick William. *Blue Water: A Tale of the Deep Sea Fishermen*. 1914; Toronto: Musson, 1935.

———. *Roving Fisherman: An Autobiography*. Gardenvale, Que.: Canadian Fisherman, 1955.

Watts, Simon, "The 1926 August Gale Haunts Aging Survivors," *National Fisherman* (Maine), vol. 60, no. 13 (Yearbook Issue 1980).

Winsor, Frederick Archibald. "'Solving a Problem': Privatizing Worker's Compensation for Nova Scotia's Offshore Fishermen, 1926–1928," in Michael Earle, ed. *Workers and the State in Twentieth Century Nova Scotia*. Fredericton: Acadiensis and Gorsebrook Research Institute, 1989.

———. "The Newfoundland Bank Fishery: Government Policies and the Struggle to Improve Bank Fishing Crews' Working, Health, and Safety Conditions, 1876–1920," PhD thesis, Memorial University of Newfoundland, 1997.

Withers, George. "French, American and Canadian Influences in the South Coast Bank Fishery," history assignment, Memorial University of Newfoundland, 2007.

IMAGE CREDITS

The Fisheries Museum of the Atlantic, Lunenburg, Nova Scotia, has generously provided most of the images for this book. The exceptions are as follows:

Nova Scotia Archives and Records Management (pages 21, 182, 183, and 193); Library and Archives Canada (pages 54, 93); Maritime Museum of the Atlantic (page iv); Burin Heritage Museums (pages 142, 154 bottom, and 155); From L.B. Jenson, *Fishermen of Nova Scotia* (Petheric Press, division of Nimbus, 1980), reproduced here with permission from his son, Lynn Jenson, Middle Musquodoboit, NS (page 144); From George Brown Goode, United States Commission of Fish and Fisheries, *The Fisheries and Fishing Industry of the United States*, Section IV (Washington 1887), (page 167).

INDEX

Fishermen who were lost and are mentioned in the text are included in this index. More names can be found in the crew lists of the lost vessels: *Sylvia Mosher*, page 78; *Sadie A. Knickle*, page 83; *Clayton W. Walters*, page 112; *Mahala*, page 116; *Uda R. Corkum*, page 118; Newfoundland vessels, page 133; *Joyce M. Smith*, page 157; *Columbia*, page 192.

A

Acadian Supplies 4, 18, 102, 103
Adams & Knickle 18, 58, 102, 114
Adventure 179, 227
Albertolite 100, 153
Alcyone 223
Alsatian 92
Anderson, Benjamin 16
Anderson, Lewis 18
Andrava 107-109, 115, 237n
Andrews, James, Reginald, and Samuel 117-118
Andrews, Wilfred 116-117, 118
Annie Anita 158-159
Annie Healy 131-133, 238n
Arleux 1, 68-69
Arras 1, 4, 100, 102, 103
Avalon 124, 191, 202, 206

B

Baker, Arthur, Blanchard, Caleb, and Guy 73, 78, 88
Baker, Enos and Titus 73-74, 78, 89
Baker, Ernest 73
Baker, Percy Amos 87
Baptiste, John 81, 83
Barkhouse, Clement 102, 103
Bay of Islands 23, 44, 172, 175
Belleoram 127, 134, 151, 178, 206
Belong, Enos 189, 190, 191, 192
Bernice Zinck 22
Blackburn, Howard 179
Blue Rocks 15, 75, 97, 102, 105, 113-114, 115, 199, 207, 212, 238n
Bluenose 21, 62-63, 102, 109, 117, 211; in international fishermen's races 180, 181-186, 192, 231;
Bluenose II 227

Bowring Brothers 201, 223
Bradley, Wendell P. 203
Brennan, Carrie and Edward 130
Bridgewater 45, 64, 76, 96
Burbridge, George and Charles 155-156, 157
Burgoyne, Enos 108
Burin 137, 141, 142, 143, 153, 154, 155, 156, 172, 190, 206; Salt Pond 154, 157
Burnt Islands 128, 132, 134
Bush, Simon T.A., Robert, and Redvis 82, 83

C

C.D. Zwicker 98
Canso 23, 42, 53, 67, 100, 172, 223, 236n
Cape Breton 8, 14, 80, 172, 174, 177, 221; 1873 storm 41-44, 47, 98, 127
Cape Broyle 24, 120, 122, 146, 172
Cashier 51, 148
Chaulk, John 128, 133
Cheeseman, Danny 131, 133
Cheeseman, Robert and Philip 156, 157
Chéticamp 80
Chiasson, Joseph, Amédée, and Cyrille 80, 83
Chute, Arthur Hunt 205, 226
Clarence T. Foote 148
Clayton W. Walters 3, 4, 82, 99, 100, 110-112, 189, 191
Cluett, Abraham Thomas 151
Coaker, William 152

248

INDEX

Columbia 192-195, 227; in August 1927 storm 102, 124, 187-192, 200, 221; in international fishermen's races 182-183; compensation 201-202
compensation for losses 200-202
Connolly, James Brendan 164-165, 166, 168, 175, 184, 185, 194, 208
Conrad, Alfred and Selborne 111, 112
Conrad, Harris 86-87
Corkum & McKean, Bridgewater 44-45
Corkum, Charles J. 79, 83, 84-85, 90
Corkum, Eric 102
Corkum, Freeman, captain 4, 102, 103, 116
Corkum, Freeman, Feltzen South 77, 78
Corkum, Leo P. 106-107
Corkum, Robert 92
Corkum, Scott 103
Covey, Arthur and Charles 97
Crann, Isaac 149

D

Day, Frank Parker 76-77
Dedrick, Frank 189, 190, 192, 202
Dennis, Clara 74, 84
Dennis, William 181
Dicks, Oliver 131, 133
Digby 35, 96, 97, 178, 223, 235n
Donald A. Creaser 205
Duff, William 64, 70, 79
Duncan Dorothy 8, 217
Duncan, Norman 216
Durnford, Arthur 132, 133
Durnford, Richard 148

E

Edith Newhall 99, 107
Effie May 132, 133
Eisenhauer, James 17, 18
Elsie 181, 186
Elsie M. Hart 29, 58, 120
Ernst, William G. 64, 100-101
Esperanto 181, 186

F

Farewell, Thomas 155-156, 157
Feener, Freeman 59

First World War 3, 33-35, 76, 141, 150-151, 197-198
Firth, Allister and Arthur 189-190, 192, 201
Firth, Lemuel R. 179
Firth, Samuel 81, 83, 190
Flat Island, Placentia Bay 130, 134, 149, 238n
Flora Alberta 238n
Foote Brothers 148
Forsey, James 51
Fortune Bay riot 172-173
Fox Harbour, Placentia Bay 131-132, 134
Fralic, Dean 98
Fraser, Chelsea 134, 162

G

General Haig 223
George Foote 51, 148
Gertrude L. Thebaud 184, 186
Getson, Daniel 50
Getson, Herbert 88, 237n
Getson, James 87
Getson, Murdock 74
Getson's Cove (LaHave) 15, 27, 50
Gladys B. Smith 207
Gladys M. Hollett 151
Golden West 87-88
Gorton-Pew Fisheries 169, 226
Graham, Leaman 75, 78
Grand Bank 141-142, 143, 147, 148, 162, 172, 218, 219, 226
Grandy, Silas 59
Greek, Currie 115
Greek, Ellsworth 115
Greek, Howard 212
Greek, Mrs Alexander 207
Greek, Rounsfell 75, 78
Green Harbour, see West Green Harbour
Gulf of St Lawrence 17, 44, 48; New England fleet in 45-46, 168, 169, 171-172

H

Haliburton, Thomas Chandler 15
Halifax 8, 13, 14, 18-19, 32-33, 95, 96, 97, 181-182, 223
Harris, Samuel 141

Hartley, Jacob 112
Harvey, C.H. 61, 67, 68, 77
Hayden, James 129
Hayden, Sterling 163, 202
Hayden, Thomas 129
Heisler, Leslie 118
Helen B. Thomas 223
Helen Miller Gould 178, 221
Hemeon, Manus 113
Henry Dowden 125, 127
Henry Ford 182, 186
Henry, H.F. 61, 64, 77, 95
Hilda Gertrude 131, 133
Hiltz, Clem 63
Hiltz, George and Reuben 118
Hiltz, Maurice 58
Himmelman Supply Company 18
Himmelman, Albert, captain 115
Himmelman, Albert, Benjamin, Eli, and George 51
Himmelman, Hastings and Nita 76, 78, 200
Himmelman, Henry and Stannage 51
Himmelman, Dr Herbert 231
Himmelman, Tommy 181
Howe, Joseph 52-53
Hynes, Leo 179
Hynick, Richard 59

I
Indian Point 107, 117-118, 199
Isnor, Lemuel 107-108

J
J.N. Rafuse & Sons, Salmon River, Digby County 109
Jacobs, Solomon 173, 178, 221
Jean Smith 58, 59
John C. Loughlin 130-131, 133
John McLean & Sons, Mahone Bay 71, 78, 114
Joyce, Albert 149
Joyce M. Smith 99, 100, 109-110, 153-157, 186, 200, 201, 239n
Julie Opp II 113, 191

K
Kearley, William A. 127

Keating, Archibald 157
Keno 115
King, W.L. Mackenzie 70-71
Kinley, J.J. 69-70
Kipling, Rudyard 164, 165
Knickle, Albert 102
Knickle, Austin 108
Knickle, Clarence 226
Knickle, Granville, Lee, Owen, and Ronald 114, 116
Knickle, Roland 79, 102, 107-108
Knickle, Warren 62, 92, 113-114, 116

L
Labrador 125-126; Lunenburg fishery 13-14, 16, 18, 26; Newfoundland fishery 36, 138-139, 145
Lady Laurier 67, 182
LaHave 10, 15, 18, 26, 44, 50, 51, 67, 80, 90, 110; LaHave Islands 34, 73, 79, 82, 87, 199, 200, 209
LaHave Outfitting Company 18
Langille, Ervin 214
Leta J. Schwartz 222
Liverpool 13, 19, 71, 79, 96, 98, 187, 202
Lockeport 81, 113, 141, 174, 223
Loughlin, Albert, Charlie, Fred, and Winnie 130-131, 133
Louisbourg 8, 70, 97, 101, 221
Lunenburg Outfitting Company 18, 79, 102, 110, 231

M
MacAskill, Wallace R. 215, 218
McFarland, Raymond 162, 176
McGill Shipyard, Shelburne 110, 223
McGuire, Agnes G. 32, 105, 106, 113
McKay, Colin 181, 224
MacLennan, Hugh 8, 217
McLeod, Robert R. 20, 176
McManus, Thomas F. 184, 222-223
Magdalen Islands 44, 45, 121, 207
Mahala 62, 92, 101-102, 108, 113-115
Mahone Bay 10, 71, 76, 78, 114, 117, 118, 204
Marian Belle Wolfe 205
Marie A. Spindler 58
Marshal Frank 99, 100, 115

Martell, Thomas 81, 83
Mary Bernice 159
Mary Ruth 86
Marystown 130, 154, 156, 158-159
Mason, Rector 108-109
Mason, Russell 114, 116
Maxner, Edward 100, 109-110, 154, 157
Maxner, William 109, 154
Maxwell F. Corkum 106
May, George Robert 205
Mayflower 186, 200
Mayo, Albert, Bert, George, and Joseph 190, 191, 192
Meisner, Frank 58
Metamora 150
Miller, Scott 114, 116
Mooween 223
Morash, Solomon 231
Morrissey, Clayton 178, 182, 202
Mosher, Allan 78
Mosher, Aubrey 73, 78, 89
Mosher, Gordon 107
Mosher, John D. 72, 74, 78, 89
Mount Pleasant 79, 81-82, 84, 90, 209
Muise, Stanislaus 80, 83
Mullins, John and Michael 131-133
Myhre, Gjert 86, 194

N
Newport, William 59
Nova Scotia Shipbuilding and Transportation Company, Liverpool 79
Noxall 127

O
Oickle, Leslie 111, 112
Ossinger, Kenneth and Warren 97
Oxner, Harry 226

P
Parks, M.J. 67-68, 69, 79, 82
Partana 92, 99
Patten, Samuel 148
Pentz, Amos 223
Peterson, John 109, 154, 157
Pierce, Ross 81, 83-84
Pike, John 156-157

Pine, Ben 102, 178, 182-184, 187-188, 191, 194, 201, 202, 221
Pittman, Jim 219
Port Hood 44, 45, 172, 174
Port Medway 4, 45, 179
Prince Edward Island 43-44, 45, 46, 50, 146, 169
Procter, George H. 45, 162, 168, 170, 175; Procter Brothers 211

Q
Quadros, Alvaro 186, 200

R
Raddall, Thomas H. 56-57, 215
Reason 149
Red Harbour, Placentia Bay 130-131, 134
Reinhardt, Kenny 34, 209-210, 212
Rencontre West 132, 134
Restless 49-50
Rhodes, Edgar Nelson 70
Richards, Fred 67
Richardson, Evelyn 215
Risser, Frank 99
Robert Esdale 78
Robin, Jones & Whitman, Lunenburg 72, 102
Robinson, Andrew 167
Romkey, Gordon Emerson 66, 199
Roué, William J. 181, 185
Rushoon, Placentia Bay 131, 134

S
Sable Island 52-57, 65, 69-70, 77, 91, 105-106, 213
Sadie A. Knickle 67-69, 79-83, 90
St. John's 48, 52, 126-127, 136, 139, 141, 148, 201
St. Pierre 49, 58, 64, 128, 132, 145, 173, 186
Schnare, Richard and Robert 59
Selig, Albert 58, 102
Selig, Guy and Raymond 111, 112
Selig, Marsden 100, 110-112
Shankle, Andrew M., and Basil 82, 83
Shelburne 13, 141, 172, 176, 201-202, 223; Shelburne County 19, 80-81, 112-113, 174, 189-190, 191

Sherman Zwicker 227
Silver Thread 74, 87
Skinner, William G. 132
Smith & Rhuland shipyard, Lunenburg 116, 181, 226, 227
Smith, Benjamin C. 207
Spindler, Israel 50-51
Spindler, Willett 58
Spinney, Lemuel 187
Stevens, David 21
Stonehurst 15, 28-29, 58, 75, 115, 119, 199
superstition 209-211, 226
Surette, Lovitt 98
Sydney 23, 42, 93, 96
Sylvia Mosher 61, 65-67, 71-79, 85-86, 89, 115, 200

T

Tancook Island 204, 214
Tanner, Angus 92
Tanner, Daniel and Willietta 29, 119-123
Tanner, Elvin Alfred, and Victor 29, 118, 120-123
Tanner, George, Irving, and Wilfred 114, 116
Tanner, Gordon 58
Tanner, Guy 92, 99
Tanner, James 75, 78
Theresa E. Connor 115, 117, 226-227, 232
Thetis 44
Thomas S. Gorton 226
Thomas, William H. 179
Thornhill, Arch 144
Thornhill, Frank 214
Three Brothers 44

U

Uda R. Corkum 4, 102-103, 115, 116-118, 123, 200, 201

V

Valena R. 132
Venosta 86, 194
Vienna 128, 132, 133
Vivian Ruth 97
Vogler's Cove 4, 44, 50, 51, 74, 89, 110-112, 199

W

W.C. Smith & Company 18, 93, 100, 102, 109
Wagner, Warren and John Eldon 74, 75, 78
Walfield, Frank Emmanuel 76, 78
Wallace, Fredrick William 20, 21, 164, 176, 178, 202, 210, 212, 213, 216; novel *Blue Water* 23, 215, 220
Walsh, James and Patrick 158-159
Walters, Angus 55, 73, 102, 109, 211, 225; spring 1926 storm 62-63; international fishermen's races 180, 181-185, 193
Walters, Stannage 110
Wamback, Parker, Wade, Walter, and William 81-82, 83, 90
Warren Samuel, Sr and Jr, and Catherine 154-155
Weaver, Forden 115
Welch, Marty 178, 181
Welcome 50-51
West Dublin 51, 74, 89
West Green Harbour 112, 189, 192, 199
Westhaver, John 68
Wharton, Lewis 187-189, 191, 192
Whynacht, Bertha, Donald, Kenneth, Ladonia, Moyle L., and Wildon 74-75, 78, 89
William C. Smith 29, 121
Williams, Charles 205
Williams, Gordon and Raymond 111, 112
Winters, Henry 102
women 13, 28, 114, 156-157, 199, 200, 201-202, 207, 218-220; Nita Himmelman 76; Florence Mosher 72; Willietta Tanner 29, 119-123; Hazel Selig 110; Catherine Warren 155; Bertha Whynacht 74-75

Y

Yarmouth 19, 20, 65, 70-71, 80, 90, 96, 176, 178

Z

Zinck, Dr Russell 231
Zwicker & Company 1, 4, 18, 59, 102, 227, 231